Representation Theorems in Computer Science

Özgür Lütfü Özçep

Representation Theorems in Computer Science

A Treatment in Logic Engineering

Springer

Özgür Lütfü Özçep
Institute of Information Systems
University of Lübeck
Lübeck, Germany

ISBN 978-3-030-25787-3 ISBN 978-3-030-25785-9 (eBook)
https://doi.org/10.1007/978-3-030-25785-9

This Springer imprint is published by the registered company Springer Nature Switzerland AG
The registered company address is: Gewerbestrasse 11, 6330 Cham, Switzerland

To my parents Kadriye and Namık Özçep

Preface

In many applications of computer science (CS), formal specifications are an important tool for the construction, the verification, and the analysis of systems. The reason is that without formal specifications one could hardly explain whether a system worked "correctly" or showed an "expected behaviour". Examples of formal specifications can be found in classical CS applications such as database applications, natural-language processing relying on knowledge bases or autonomous systems (agents) which require formal specifications for goals, plans, or hypotheses on the current state of the environment.

A system designer faces the challenging task of designing formal specifications that capture his intended model of the application domain or the intended behaviour of the system as appropriately as possible. But the chosen formal specification may fail to capture the intended model or may contain non-intended models. The latter happens, in particular, in those cases where the expressiveness-feasibility balance was chosen in favour of a polynomial-time algorithm instead of an expressive language. Testing for the existence of non-intended models from a semantical perspective is quite more complex than a satisfiability test of the formal specification.

The missing quality feedback during the design of formal specifications in the area of databases, knowledge bases, and agents is the motivation for a theoretically important and practically relevant research question: given a formal language from one of the applications mentioned above and given a formal specification in that language, how can one develop a deep understanding of all models of the specification? In following this research query, the work described in this monograph relies on a method based on so-called representation theorems. A valid representation theorem for a formal specification states that the set of all models of a formal specification is representable by a subset of all models. Hereby, the representing models are required to be constructed according to a simple principle.

This monograph develops representation results in the context of three applications mentioned in the beginning. The first is from the area of qualitative spatial reasoning and considers binary relations of spatial relatedness. Human perception of spatial relatedness depends on the type (granularity) of the considered objects: for example, when answering the query asking for all objects near a tree, a different scaling context is relevant compared to the case where one has to answer the query asking for all objects near the city of Lübeck. This monograph develops a model of spatial relatedness that captures the scaling context with hierarchical partitions of a spatial domain, and characterises the resulting relations axiomatically.

The second case example is stream data processing. In this monograph, streams are modelled as potentially infinite words, and stream processing is abstractly modelled by functions that map input streams to output streams. This monograph shows that various important properties of stream processing such as prefix-determinedness or various factorisation properties can be axiomatised, and it further shows that the axioms are fulfilled by natural classes of stream functions.

The third case example is belief revision, which is concerned with the revision of knowledge bases under new, potentially incompatible information. In this monograph I consider a subclass of revision operators, namely the class of reinterpretation operators, and characterise them axiomatically. Reinterpretation operators dissolve potential inconsistencies by reinterpreting symbols of the knowledge base.

The concept of representation can be identified also in the paradigm of ontology-based data access. In strict ontology-based data access, the problem of query answering is reduced to answering a query over the data. To this purpose the original query is rewritten such that the intensional knowledge of the ontology is captured by the rewritten query. However, rewriting is not possible for all ontology and query languages—and the proofs in case of rewritability are not trivial. Rewritability can be proved with so-called universal models—and this is where the representation aspect is relevant: universal models are models that represent all models of an ontology.

In this work, mainly two rewritability results are shown: In the area of qualitative spatial reasoning it often happens that one has to answer queries over complex concepts that combine thematic and spatial aspects. This monograph develops various potential ontology languages and shows for which rewritability is given. In the area of stream processing a rewritability result is given for the stream-temporal query language STARQL.

The results of this monograph demonstrate the use of representation theorems for the design as well as the evaluation of formal specifications and, hence, they can be conceived as theoretical contributions for building future application-development kits that support application designers with automatically built representations.

Most of the results presented here were published within the last seven years during my habilitation and were presented in various venues and to various research communities from AI, description logics, semantic web, geographic information systems, and databases. This monograph presents the contents of the papers from the perspective of representation and gives full proofs, which, due to page restrictions, were either sketched only or were not given at all in the papers. For the sake of readability I decided to present the proofs in chapter-wise appendices.

Lübeck, *Özgür Lütfü Özçep*
April 2019

Acknowledgements

I would like to thank Prof. Dr. Ralf Möller for all his support and engagement during the habilitation phase. It was a great pleasure to discuss and work with him on the topics of this monograph, which would not have been written without his enduring motivation and patience, his thoughtful comments, and his many helpful ideas.

Usually, when he had shared one of his ideas with me, he added in an inimitable humorous style: "Jetzt musst Du es nur noch LaTeXen!" ("Now you just have to LaTeX it!"). Of course there was much more work than what these words entail, so I take full responsibility for all glitches and errors that might have crept in.

I also would like to thank the three reviewers of my habilitation thesis, which served as the basis for this monograph.

Last but not least let me express my deepest sincere gratitude to my family for all the support of the non-academic kind.

Contents

Chapter 1
Introduction

Abstract In many applications of computer science (CS), formal specifications are an important and even necessary tool for the construction, verification, and analysis of systems. The introductory chapter illustrates this observation with three classes of applications that a computer scientist might be familiar with: databases, knowledge bases, and agents. Moreover, it discusses the role of logic in such applications, states the problem of having possibly many unintended models of formal specifications, and sketches the proposed solution based on representation theorems.

1.1 Motivating Examples

Three typical applications in which formal specifications are important are given in the following subsections.

1.1.1 Databases

In database applications, formal specifications appear, e.g., in the design of schemata for data representation. Every student of database systems is taught to build entity-relationship (ER) models or UML diagrams, and—based on these—to design appropriate database tables. These formal specification frameworks allow for describing the relevant entities of a given domain and also the relations that hold between them. On the database level the ER model is reflected in the choice of tables, columns, and integrity constraints such as primary and foreign keys to which the data have to adhere. Data are accessed via declarative query languages such as SQL for which the intended meanings of queries, i.e., the intended sets of answers, are specified formally.

Chapters 4 and 5 of this monograph are related to some aspects of formal specifications in databases.

Ö. L. Özçep, *Representation Theorems in Computer Science*,
https://doi.org/10.1007/978-3-030-25785-9_1

1.1.2 Knowledge Bases

Many applications that are based on processing natural language (NLP) or controlled natural language (CNL) rely on the use of *knowledge bases* which represent terminological knowledge and facts via logical formulae. Query answering (QA) over knowledge bases is quite more challenging than QA over databases. The reason is that, in order to ensure the correctness and the completeness of the set of answers, implicit knowledge following from the logical formulae has to be taken into account. The research described in Chap. 3 and Chap. 4 of this monograph is motivated by the design of knowledge bases that can be used, e.g., in geographic information systems (GIS) in order to store spatio-thematic objects and access them via queries such as "Show me all secure playgrounds that are nearby!".

An equally challenging task as QA is that of storing new data in knowledge bases. The reason is that new data may not be consistent with the knowledge base at hand, so that a revision of the knowledge base is required in order to ensure its consistency (similar to ensuring integrity of databases). *Belief revision* is a field in the intersection of CS, logic, and theory of science providing a general strategy to deal with potential inconsistencies under new information. A specific form of belief revision that is useful for applications such as ontology development, ontology debugging, and ontology alignment is the topic of Chap. 7 of this monograph.

1.1.3 Agents

In systems with autonomous sub-systems, alias agents, even further formal specifications are required, say, in order to describe the intended behaviour of agents in their dynamic environment. A rational agent perceives sensor data from an environment and acts therein so "as to achieve one's goals, given one's beliefs" [30, p. 7]. This simple yet fundamental characterisation hints to the challenging problem of transforming sensor data into high-level conceptualisations—required for reasoning with beliefs and goals—and transforming operations on the higher level to operations on the lower level. Hence, for the design of agents, formal specifications of the following kinds are needed: specifications of the background knowledge in a knowledge base, specifications of goals, possible actions, and plans as well as specifications of beliefs on the current state of the environment. Needless to emphasise that the kinds of transformations should be performant because agents acting in a dynamic environment perceive information via (possibly high-paced) streams of timestamped data, and because not all data elements can be stored. In Chap. 5 of this monograph, foundational aspects of stream processing are discussed. In Chap. 6 the focus is on high-level declarative stream processing with a query language enabling QA w.r.t. a knowledge base.

Computer science draws its tools for formal specifications from mathematical logic, a field studying different types of specific logics[1]. In fact, in all CS applications mentioned above, the formal specifications are either specific formulae in a specific logic or at least related to logical formulae. The following section describes the role of logics in more detail.

1.2 Role of Logics

Ever since the rise of CS as a fully fledged discipline on its own, it had—and continues to have—fruitful interactions with the field of logic. The influence of logic on CS is considered to be that strong that some researchers talk of the "Unusual Effectiveness of Logic in Computer Science"[2], comparing it with the "Unreasonable effectiveness of Mathematics in the Natural Sciences" [36]. The authors of [12] even go further and consider the birth of CS as an outcome of the development of the field logic[3], in particular as outcome of the ambitious automatisation program for mathematics conducted by Hilbert and colleagues from 1900 to 1928.

That logic cannot serve as the foundational framework for mathematics—as envisioned by Hilbert—was proven with results of Gödel on the incompleteness of arithmetics and on the non-provability of set theory in between 1931 and 1933 as well as with results of Church and Turing in 1936/37 on the non-decidability of validity for first-order logic (FOL). The awakening caused by these results also led to a shift in focus, away from problems of pure mathematical logic to CS-related aspects of logics—indicating the impact of CS to logic: algorithmic aspects of important problems from logic, such as model checking, validity/satisfiability checking etc. have become important research topics in logic. And also here the impact is very deep. A point in case is the fact that there is a plethora of "industrial logics" (to borrow a term from Vardi's paper [34]). As the term indicates, these kinds of logics are really used for practical problems arising from industrial needs and as such, roughly, share the property of an adequate balance between expressivity and feasibility. The challenge in finding the right balance should be obvious: solving algorithmic problems defined over specifications in a more expressive logic is not feasible for certain instances, i.e., will require non-polynomial time or more than logarithmic space. And so in each individual CS application, the application designer has to achieve the right balance.

Fortunately, in searching for an appropriate logic with the required complexity, system and application designers can rely on theoretical results from the field of descriptive complexity [13, 14]. Descriptive complexity investigates correspondences between the expressivity of a logic and the complexity of solving problems formu-

[1] Note the plural use: there is the field of logic having various logics as its objects of study.

[2] This is the title of a paper which appeared in the Bulletin of the Journal of Symbolic Logic [12].

[3] "In the beginning, there was Logic" and "Computer science started as Logic" is the literal wording that can be found in the accompanying slides to paper [12]. See: `https://people.cs.umass.edu/~immerman/pub/cstb.pdf` (accessed: 26-12-2016)

lated in this logic. This field—and the wider field of finite model theory—started with a result of Fagin [8] that characterises existential second-order logic by the complexity class of problems solvable in nondeterministic polynomial time (NPTIME). Further correspondences were discovered soon afterwards, and actually one of them, the correspondence of the complexity class AC^0 and FOL is used also in Chap. 4 of this monograph. These correspondences give general resource bounds on what can be expressed with a given logic. For example, if a problem specified in a new language cannot be expressed in FOL, then it will require more time/space resources as allowed according to AC^0. Stated in the other direction, the AC^0-FOL correspondence says: if computing a problem needs more time/space than specified by AC^0, then it cannot be expressed in FOL. Interestingly enough, the logical characterisation of the class of polynomial time (PTIME) algorithms on finite non-ordered structures is still open.[4]

As for the reasons of the unusual effectiveness of logic in CS, the authors of [12] mention that logics have the following properties making them suitable for CS: First, a logic provides a formalism (semantics) for describing mathematical structures. In logic, mathematical structures are defined w.r.t. a signature (sometimes also called vocabulary) consisting of sets of individual names, relation names with given arities, and function names with given arities. Then, a mathematical structure for a signature consists of a domain of objects, and a denotation function that maps each individual name to an object of the domain, each relation of arity n to an n-ary relation over the domain and each n-ary function symbol to an n-ary (total) function over the domain. A simple example are graph structures. The signature consists of a single binary relation symbol E, and a graph structure for such a signature is made up by the set of vertices as the domain and a denotation for E, which is a binary relation standing for the edge relation. But also "dynamical" structures such as streams, sequences of states, or processes can be described as mathematical structures.

Secondly, a logic provides a language (syntax) based on the signature in order to describe the properties of mathematical structures. With such a language logical expressions such as terms—denoting objects of the application domain—and sentences—describing the application domain and evaluated to true or false—can be constructed. For example, in the case of graphs it is enough to have—next to logical symbols—the binary relation symbol E of the signature in order to express, say, the property of being symmetric: for all vertices x, y, if there is an edge from x to y, then there is an edge from y to x, formally: $\forall x \forall y (E(x, y) \rightarrow E(y, x))$ holds.

Related to these points, and even founding them, is a clear distinction between syntax, which governs the rules to set up the sequences of characters to talk about the objects of interest, and semantics, which governs the rules to associated (intended) meanings with the sequences. This last point regarding the distinction between syntax and semantics is of immense importance because it allows for speaking about the "intended meaning" of an expression, of its "expressive power", of the

[4] In fact, if one could prove that there cannot be a logic characterising PTIME over arbitrary finite structures, as conjectured by Gurevich, the long-standing open problem whether PTIME = NPTIME would have to be answered negatively, because—as mentioned in the text—NPTIME is characterised by second-order logic. For more details see, e.g., [16, Sect. 10.7]

"correctness and completeness" of a calculus etc. In particular, as stated above, one can define syntactical entities called *sentences* and define precisely the truth conditions of sentences, i.e., the conditions under which a structure makes a sentence true. Structures that make a sentence or a set of sentences true are called *models*. Coming back to the example applications from the beginning, it is clear that semantics formally ground and justify the notions of "correctness and completeness" of answers of an SQL query, or of the "inconsistency" of a knowledge base, or the "expected behaviour" of an agent.

Sometimes the syntax/semantics distinction is blurred, for example the construction of the Herbrand structure relies on the syntactical elements of the given language (or theory). Also, some hard-core proof theorists do not even believe in the usefulness of this distinction—at least when considering truth-conditional semantics à la Tarski.[5] Furthermore, the classical approach to belief revision [1], named AGM belief revision after its founders Alchourrón, Gärdenfors and Makinson, relies on an abstract notion of logic based on *consequence operators* (see also Chap. 7 of this monograph). These do not refer to mathematical structures, truth values or any other semantical notion at all. Nonetheless, later work on belief revision works also with the classical approach to logic. And even in the AGM approach there is a benefit of associating meanings with syntactical entities. Moreover, it is possible to simulate semantical objects on the base of syntactical entities (via maximally consistent sets). Both views on a logic—semantic-based and the one based on consequence operators—are considered as useful for the treatment of belief-revision operators. Summing up, the syntax/semantics distinction is useful, it is part of textbook definitions of logics, and it has been the base for many research questions in the field of logic.

Regarding above-mentioned observations on the common history of CS and the field of logic it does not come as a surprise that much of the current work in computational engineering involves *"logic engineering"* [2]. The monograph at hand has to be considered as a contribution to the foundations of logic engineering in various fields of CS, mainly within the larger field of knowledge based representation and reasoning, in particular within belief revision, qualitative spatial and temporal reasoning, agent theory, stream processing/reasoning, and databases. This monograph also contains results that are relevant for research on "semantic technologies" and their use in industry. However, as explicated in the following section, my insights are not the result of the current, established practice of just applying "semantical technologies" but, rather, are the result of a critical analysis of this practice.

---

[5] A radical example for abandoning the classical distinction between syntax and truth semantics is the program that Girard developed in several papers under the title "The Geometry of Interaction". In the last of these papers, entitled "Geometry of Interaction V: Logic in the hyperfinite factor" (fully revised version (October 2009)), which is an unpublished manuscript available online, he calls for a foundation of logic where the semantical objects are not primitives.

1.3 Problem: Need for Representations

Summing up the last section, in my view, logics provide the right tools and results for using formal specifications within CS applications such as those mentioned in the beginning of this chapter. But, now, the application/system designer really has to use them: he faces the challenging task of designing formal specifications that capture his intended model of the application domain or the indented behaviour of the system as appropriately as possible. But the chosen formal specification may fail to capture the intended model or may contain non-intended models. The latter happens, in particular, in those cases where the expressiveness-feasibility balance was chosen in favor of a polynomial-time algorithm instead of an expressive language.

Even if the designer decides to use a very expressive language for the formal specification, he has to take a closer look into the models of a formal specification. Because non-intended models are lurking in the guise of non-standard models in any formal specification. This holds in particular for logics for which an upwards Löwenheim-Skolem theorem can be proved. The Löwenheim-Skolem theorem also holds for FOL and it states in this case that any FOL theory having a model has a model of arbitrary size. An illustrative example [4, p. 21] is the task of modelling with FOL a discrete, infinite flow of time as a structure $(T, <)$ where T stands for a set of time points and the binary relation $<$ stands for the order of time points. In FOL it is possible to formulate a finite set of sentences DI (standing for **di**screteness) that describes the intended structure. Concretely, DI states that $<$ is transitive, irreflexive, linear, infinite into the past as well as into the future, and discrete. The designer's intended (standard) structure is the structure $(\mathbb{Z}, <_{\mathbb{Z}})$ consisting of the integers and its natural order. But in addition to this model, DI has many non-standard models such as the structure $(\mathbb{Z} \oplus \mathbb{Z}, <_{\mathbb{Z}\oplus\mathbb{Z}})$ consisting of two disjoint copies of the integers, where the order within the two instances is as in $\mathbb{Z}$ and where all elements from the first instance come before all elements of the second instance.

The existence of many (un)-intended models is not a result of inventing an artificial scenario, rather it is a typical phenomenon due to the nature of the task an application/system designer faces: he has to transform his ideas, his cognitive models, and possibly also the requirements of others—which may be formulated in natural language—into formal mathematical specifications. Hence, there is a clear need for supporting system designers with quality feedbacks for the formal specifications they construct.

The observation on the lack of a quality feedback may be at odds with the frequently claimed success of "semantic technologies" that were developed within the semantic-web and description-logics community. For sure, there are relevant theoretical results on different fragments of description logics, and there are relevant practical results regarding reasoning systems as well as standardisation issues such as the web layer stack (RDF, OWL etc.). Also, regarding the design of ontologies, state-of-the art reasoners give quality feedbacks w.r.t. simple metrics about, for instance, the number of axioms, depth of concept nesting, or semantical aspects, such as the consistency of an ontology. Furthermore, there are some elaborated frameworks for various specific tasks of ontology evaluation [35]. But all these results and existing systems should not obscure the fact of the necessity for further quality feedback for the ontology engineer. The point is that the kind of feedback mentioned above is still far from giving the designer the necessary insight into the models of his ontology. And, hence, I claim that formalisms and systems are not really used appropriately in practice. Similar claims were made some time ago by other researchers working on formal ontologies in informations systems (see, e.g., [33]).

This monograph is a contribution to filling the gap of missing quality feedbacks. It gives theoretical results on possible models of formal specifications by exploiting the mathematically well-founded idea of representation theorems, as used, e.g., in the well-known representation theorem of Stone for Boolean algebras [32]. Informally, a valid representation theorem for a formal specification states that the set of all models of a formal specification is representable by a subset of all models. The models of this subset cover all relevant aspects required to describe all models because for each model there is a structure preserving mapping from some model of the representing subset. Moreover, the models in the representing subset are constructed according to a simple principle. In the case of Stone's theorem set algebras are proved to represent the class of all Boolean algebras. (For a formal statement please see Chap. 2.)

So, how do representation theorems help w.r.t. building "intended models"? Though in many cases one cannot circumvent unintended models (as the above discussion on the existence of non-standard models showed), at least, one can "tame" them, because, each model of a formal specification, be it intended or unintended, will have a model representing it. There will be harmless unintended models in the following sense: they were not expected (they are non-standard) but they do not influence the reasoning services over the formal specifications. However, there might as well be really unintended models which one wants to exclude as they influence the reasoning services. Representation theorems explicate a possible intended hidden structure of formal specifications (see also the comments on hidden structures at the end of the next section). If the set of really unintended models is not empty, then one has at least a means to characterise them also in terms of the instances of the hidden structure—and thereby, possibly, a means to exclude them by adapting the hidden structure.

The term "representation" (and, more specifically, the term "representation theorem") is used with slightly different meanings in various fields—even within CS. But any of the different readings cover the core meaning of "representation" as used in natural language: There are two classes of objects, the first one standing for the things to be represented (the representandum) and the second one for the objects that represent (representans), and there are structure preserving mappings between the representing and the represented objects. The mappings between a representing structure and a represented structure may be different in each case, but for a given representation theorem each adheres to some constraint that ensures the property of being structure preserving.

In all of the examples considered in this monograph, the representing objects also represent themselves, so I consider both the represented and the representing objects as comprising a class of objects described as class of objects fulfilling some property (say, being models fulfilling a theory). As stated above, in any of the representation theorems the class of representing objects fulfils some form of homogeneity in the sense that all representing objects can be described by some property or by some construction principle. Figure 1.1 illustrates the general representation scenario.

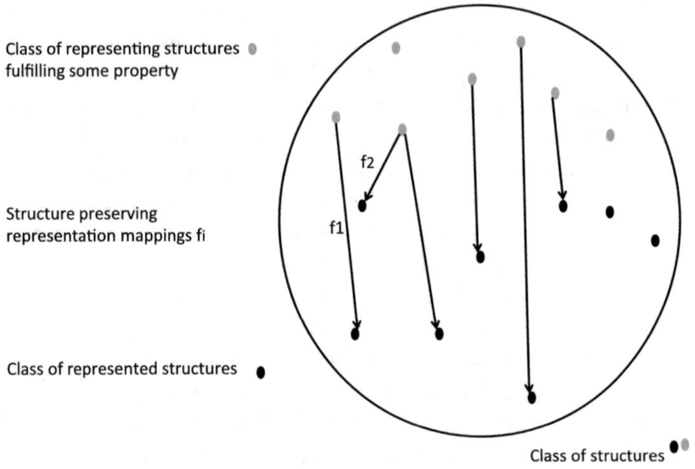

Fig. 1.1 Components of a Representation Scenario

The representation theorems proved in this monograph can be considered as a contribution to a necessary development in (industrial) applications of CS, away from blind trust in abstract semantics of formal specifications towards a better—in fact: feedback supported—understanding of the concrete semantics of formal specifications. This will give general (industrial) standardisation aims regarding representation structures the necessary justification.

In order to illustrate the point of the last paragraph, a simple DB design scenario for building a university administration system might be helpful. Assume that the designer has to model the fact that all professors are employees of a university. Also he has to model the fact that each university is an employer of, amongst others, professors. It is clear, that the relation of being an employee of a university is the inverse of the relation of being an employer. The DB designer can use the machinery of integrity constraints—provided by any database management system (DBMS)—in order to express the inverse property. This will guarantee that in each concrete DB implementation the following will be ensured: if there is an entry (a row) stating that a professor is an employee of a university, then there will be also an entry stating that the university is the employer of the professor. On each DB this integrity constraint can be checked by the DBMS. But such an inverse constraint is extensional. It actually means describing data structures that should hold in any concrete DB with concretely named elements. The general idea that the relations employee-of and employer-of are inverse is not represented intensionally. To give a more drastic example: in the concrete world it might be the case that all humans are exactly those beings with an earlobe. But this constraint does not represent the real essence of a human being, as one could think of a possible world in which humans do not have earlobes.

Ontologies mitigate the problem of a lacking means for intensional representations. In the university scenario, the ontology could be designed to contain an axiom stating that the relation employee-of is the inverse of the relation employer-of. This axiom constrains any model so that the following holds: if the pair (professor, university) stands in employee-of relation, then the inverse pair (university, professor) stands in employer-of relation. And this holds also for pairs of objects that are not even named by some pair of constants within the model. Having the ability to talk about objects that are not named means a big and necessary step upwards in expressivity. The reason is that there might be things, say abstract things, that are not named but which should be assumed to exist in order to explain some or other observation within the concrete application domain.

But how does the designer control those entities which are not named? How can he be sure that the logical constructs of the ontology (the inverse operator, the logical constructors, the signature he has chosen) are sufficient to constrain the models? How does he tame also the non-named objects, the concepts, and relations in which they may participate? There are ontology reasoners which do the analog of integrity checking provided by DB management systems, namely, checking for the consistency of an ontology, i.e., checking whether there is a model at all. But due to the intensional flavour of the axioms, which allow for expressing constraints also on non-named objects, the space of possible models appears to be quite more diverse. In fact, the kinds of possible constellations between the entities are affine to, sometimes, surprising (alias non-standard) constellations. Hence, one needs more than the service of consistency checking. One needs a guide on how to explore the space of possible models. Searching for representation theorems is an appropriate methodology. (For a different methodology that was developed in a software-engineering context please see the Alloy language and the Alloy-based support tools described in [15].)

1.4 Contribution: Representation Theorems

This monograph develops representation results similar to that of Stone's representation theorem within three different fields, the field of qualitative spatial reasoning, the field of stream reasoning, and the field of belief revision. Moreover, this monograph presents further results that are related to representational aspects, namely rewritability results in the context of the paradigm of ontology-based data access (OBDA). I am going to sketch the results in more detail in the following paragraphs.

Most of the results presented here were published within the last seven years [24, 19, 26, 20, 21, 22, 28, 27, 23, 17, 25, 18] and were presented in various venues and to various research communities from AI, description logics, semantic web, GIS, and databases. All papers contributed to current research topics but already contained, though rather latently, aspects of representation. This monograph presents the contents of the papers from the perspective of representation and gives full proofs, which, due to page restrictions, were either sketched only or not given at all in the papers. The hot-topics of the last seven years touched with the above-mentioned papers might be considered outdated in the future, but the general aspects of representation underlying them are so fundamental that I expect them to have a long-term effect.

The first representation result described in this monograph is developed in the broader context of qualitative spatial reasoning (QSR). The focus of the result lies on binary relations of spatial relatedness—measuring the distance of two sets over a given set X. So, the underlying signature σ_{QSR} is mainly that of set theory together with a binary symbol δ, expressing spatial relatedness, and the structures are of the form $\mathfrak{A} = (pow(X), \delta^{\mathfrak{A}})$.[6] I consider a set of axioms Ax over the signature σ_{QSR} which describes some desirable properties for the relation δ. As domains of the considered structures are of the special form $pow(X)$, the structure preserving condition is actually that of identity. So the representation theorem boils down to the following: for a given set of axioms Ax find a class of σ_{QSR} structures Y_{Ax} such that for any σ_{QSR}-structure $\mathfrak{A} = (pow(X), \delta^{\mathfrak{A}})$ fulfilling the axioms one can find a structure $\mathfrak{A}' = (pow(X), \delta^{\mathfrak{A}'}) \in Y_{Ax}$ such that $\delta^{\mathfrak{A}} = \delta^{\mathfrak{A}'}$.

Of course this form of representation becomes interesting only if the set Y_{Ax} consists of "similar" structures which have the same construction principle (as they result from instantiating a set of parameters, say) or which can be characterised by a common property. Indeed, in case of Stone's representation theorem the common property of the class of sets of fields is that they share a specific domain (the power domain) and that the operators are concrete operators on sets. So, in this case, the common property is circumscribed by a set of axioms of set theory which allows for defining the notion of a power set.

The main result w.r.t. binary relations of spatial relatedness is the axiomatic characterisation of a special class of binary relations for spatial relatedness. In this class the type of an object determines the scaling context under which to measure spatial relatedness. For example, deciding spatial relatedness w.r.t. a district has to be conducted in a different scaling context than for deciding spatial relatedness w.r.t. a city or w.r.t. a country.

The second thematic area for which this monograph gives a representation result is that of stream processing and stream reasoning. Streams are potentially infinite sequences of elements. Streams of different types appear in various applications of CS such as sensor networks, data-stream management systems, complex event processing, or autonomous systems (agents) which—in addition to streams of percepts—have to process high-level data such as actions, goals, beliefs etc. The main challenging aspect of stream processing is the task of continuously processing data without storing every bit of them (because there is potentially too much to be stored) and without falling behind.

In case of stream reasoning the logical objects of the representation study are functions with streams as input and output. In this monograph, I describe repre-

[6] Please note that I have left out all set-theoretic relations, functions and constants as these are assumed to have a fixed interpretation according to set theoretical axioms.

sentation results for stream functions based on the work of [11], which considers streams from the infinite-word perspective. The main observation of [11] can be stated as a representation result: every stream function that is prefix-determined can be represented as a function that is constructed as an iterative application of a sliding window over the stream. In this monograph I generalise this observation in order to characterise stream functions based on different types of window functions with factorisation properties. An especially interesting class of windows regarding feasibility are bounded-memory windows, i.e., window functions that can be evaluated using constant space only.[7] I give a purely functional description of this class using a set of initial functions and the principle of recursion.

The most natural type of stream data are one-dimensional temporal data, i.e., data with a tag for a time point. But also many-dimensional temporal data and, even more general, also non-temporal data can be processed in stream-wise manner—as long as a procedure for sequentialising the data exists. Hence, stream processing is also relevant for the area of qualitative spatial reasoning. For example, if one is interested in answering queries over 2-dimensional rectangular regions, then one can access these regions in a stream-wise manner by applying a small squared window function that is slit over the regions from, say, left to right and from top to bottom.

As a third area for a representation result in the style of Stone I consider belief revision. Here the technical objects of interests are structures consisting of a set-theoretic domain and a binary operator. Belief revision deals with the dynamics of declaratively specified repositories (say DBs or knowledge bases) where the change is triggered by new incoming information that has to be inserted into the repository (revision) or which has to be deleted from the repository (contraction). So belief revision considers binary change operators which, for a given pair of repository and trigger information, output a new repository. The main representation result of this monograph within the context of belief revision is the characterisation of a specific class of revision operators that are meant to dissolve ambiguities. Sometimes the inconsistency between the trigger and the knowledge base is not due to previously obtained false information in the knowledge base but due to different uses of the same terms by the holder of the knowledge base and the sender of the trigger. For example, the trigger sender may have a more strict understanding of "article" (an article must be published in a journal) than the holder of the knowledge base (an article can be published either in the proceeding of a conference or in a journal). The ambiguity is dissolved by reinterpreting the use of "article" in the knowledge base in favour of the reading according to the trigger.

If one has identified a representing system of objects for a given class of objects, one has already everything at hand that characterises this class. Nonetheless, it may be the case that there is not only one construction principle and hence not only one class of representing objects but more than one. It is quite instructive to consider different representation systems for a class of objects as one may have further properties not shared by the other system—say, one of them might be more feasible than the other, or might be better suited for implementation etc. Applying this idea to belief

[7] "Windows" in the sense of this monograph are allowed to be unbound to the left.

revision, one can think of the existence of more than one construction principle for the same revision operator: the concrete operators have the same input-output behaviour but are constructed according to different principles. An example in case of such a situation in belief revision is the identity of the skeptical operators of Delgrande and Schaub [7], which use bridging axioms, and the revision operator of Satoh [31], which uses a minimal distance approach. I discuss this identity in Chap. 7. In that chapter the idea is pushed even further: I do not search for representing belief revision operators for a given set of operators of an axiom system, but, rather, I try to show how to simulate belief revision operators adhering to some construction principle with belief revision operators adhering to the construction principle of reinterpretation.

In all three representation results the representation aspect is clearly identifiable. As mentioned above, in this monograph I also present results in the area of OBDA for which the representational aspect is not that obvious but clearly existent. Because it is rather unusual to consider OBDA and in particular its core technique of query rewriting from the perspective of representation, I am going to sketch in the following the representation aspects of OBDA in more detail, anticipating the technical terminology introduced in Chap. 2.

In OBDA, data stored in databases are accessed via an ontology. An ontology represents terminological knowledge. Using the distinction of descriptions logics, an ontology in the narrow sense is referred to as tbox (for terminological box). And the data are represented as facts in a so-called abox (assertional box). The user can query the data via a query language that uses the vocabulary of the ontology. Query answering, as stated in the beginning for applications on knowledge bases, is difficult, as usually answering the queries requires to take into account the implicitly derivable knowledge of the ontology.

For example, assume that the ontology models the administrative structure of a university and has, amongst others, an axiom stating that all PhD students are students. If in the database Peter is recorded as a PhD student, then together with this axiom the implicit assertion that Peter is a student can be derived. In particular, this means that the set of answers to a query that asks for all students has also to contain Peter, though the database does not explicitly mention Peter as a student. This example illustrates the so-called *certain semantics* for query answering: infer all and only those knowledge that is certain w.r.t. the tbox and the abox. Using the notion of models this amounts to: find the answers that hold in all possible models of the tbox and the abox.

In OBDA in the classical, stricter sense (sOBDA) the task of computing certain answers is done via rewriting: the reasoning service of query answering w.r.t. a tbox is reduced to answering a rewritten query (expressed in FOL) over a database associated with the given abox. So, instead of using a reasoning system which deduces the answers to the query, the query answering service is reduced to the comparatively simple[8] service of query evaluation on a database. But query rewriting is not always possible. In order to guarantee it, strict expressivity constraints have to be obeyed by

[8] Here "comparatively simple" means simple w.r.t. the complexity class hierarchy: answering FOL queries on DBs is in AC^0 w.r.t. data complexity (see below).

the ontology language as well as the query language. Finding the right balance of these languages is quite challenging.

The results in Chapters 4 and 6 are FOL rewriting results in the sense explained above. I show FOL rewritability for a specific logic in the realm of QSR and, secondly, for a logic within the realm of stream processing. But how is FOL rewritability related to the notion of representation? One way of showing that rewritability is possible is via a specific model of the tbox and the abox, namely a universal model. Universal models are an important construction in OBDA but also in other areas such as database theory or data exchange [9, 3]. The most general definition of the property of universality can be given within the abstract framework of category theory (see any book on category theory, such as [10]). But for the purposes of this monograph it is sufficient to stick to the concrete notion of a universal structure in terms of model theory. It says that a model is universal if and only if for any model of the theory (tbox plus abox) there is a structure preserving mapping, technically: a homomorphism, from the universal model into it. And this is exactly the scenario of representation where the class of structures to be represented is the set of models of the tbox and the abox and where the representing class consists of the chosen universal model. Rewriting a query amounts to finding a new query by which the universal model is captured.

As mentioned above, the main aim of this monograph is to fill the quality feedback gap for application and system designers by helping them getting a deeper understanding of the properties of their formal specifications. This approach has also the nice side-effect of mitigating a tendency in current practice of ontology engineering as explained in the following. In the early days of description logics (DLs), the tbox (terminological box) was defined to consist of axioms of a very simple kind, namely explicit definitions. A *DL explicit definition* is an equivalence axiom stating that a concept to be defined, the definiendum, is equivalent to a concept with which one defines, the definiens. The motivation for relying on such restricted forms of axioms is the observation on a natural distinction of expressions: Firstly, there are primitive expressions for which one does not stipulate any definition at all as these are considered to be atomic or basic. And secondly, there are non-atomic, namely defined expressions which are constructed explicitly via primitive expressions. Explicit definitions fix the meaning of the definiendum uniquely in any model—given the denotations of the primitive symbols.

In the current practice of ontology engineering with DLs, instead of explicit definitions, arbitrary axioms, so-called general inclusion axioms (GCI), are used. These still might lead to *implicit definitions* in the following sense: it might be the case that in any model of the GCIs the extension of the non-primitive concepts are already uniquely determined by the primitive concepts. But this is not guaranteed[9], and hence even the distinction of primitive and non-primitive concepts is not part of the current ontology engineering practice. With the representation theorems the ontology designer gets back this distinction in a different disguise: He stipulates some axioms with expressions that he thinks denote the relevant entities of his domain. If

[9] See [6] for a discussion of this point w.r.t. expressive description logics.

a representation theorem holds, then the representing models provide the primitive entities that are considered to be relevant for the application domain.

The search for representation theorems can also be seen as search for *hidden* or *latent structures*. The general idea of hidden structures is that there are features that can be observed and others for which one does not have observations but which are useful in order to explain or to generate the observed features. This conception of hidden structures is known to most computer scientist from the area of machine learning but is inherent also to the representation idea explained above. For example, in order to describe some properties of a system the designer sets up a theory using a vocabulary consisting of such constants, relations, and functions that are required to denote the entities of the domain. In order to represent all models, the designer has to search for the hidden structures that generate (or represent) the models. The hidden structures may rely on a completely different vocabulary not mentioned in the vocabulary for specifying the theory. To illustrate this point: In case of spatial relatedness discussed in Chap. 3 the models to be captured are binary relations fulfilling a specific set of properties. The hidden structure generating such relations are hierarchical nestings of sets (called partition chains, see below). In case of operators on streams (as discussed in Chap. 5), the hidden structures are window functions. In case of belief revision (as discussed in Chap. 7) a possible hidden structure is the set of bridging axioms of a specific kind.

1.5 Overview of Chapters

Chapter 2 provides necessary terminology and knowledge that is used throughout this monograph. In particular, it introduces necessary notions for working with logics, their syntax and semantics, gives an overview of description logics (amongst others the family of lightweight logics DL-Lite), and introduces relevant notions for the paradigm of OBDA. Based on these, the notion of representation as used in this monograph is stated more formally. Also, this chapter contains a short introduction to the region connection calculus (RCC), which is considered in Chap. 3 and in Chap. 4.

Chapter 3 discusses my main representation result in the area of QSR. The objects of interest are binary relations of spatial relatedness (nearness) which capture the frequently observable fact that nearness is determined by a scaling context. This scaling context can be modelled by a tree-like structure called a nested partitioned chain where the root describes the roughest scale and the leaves describe the finest scale. In the first part of the chapter I consider the arguments of the spatial relatedness to be (ordinary) sets over a given finite set. In the later parts I discuss two extensions: 1. How to track the changes in spatial relatedness when the underlying partition chain changes. 2. How to deal with spatial relatedness when the arguments of spatial relatedness are not ordinary sets but regions according to the region connection calculus.

Chapter 4 is also a contribution in the realm of QSR. The chapter discusses ontology and query languages that are meant to provide access to spatio-thematic objects. Because in most geo-processing scenarios one has to cope with large data volumes, I explore combinations of description logics with qualitative spatial calculi that could be potential candidates for an OBDA approach. In particular, I show that a weak combination of DL-Lite [5] with the most expressive region connection calculus RCC8 [29] allows for FOL rewriting whereas for stronger combinations of DL-Lite even with less expressive calculi such as RCC3 and RCC5 lead to non-FOL rewritability.

Chapters 5 and 6 are contributions to stream reasoning, Chap. 5 having a more foundational character and being relevant for any type of stream reasoning, and Chap. 6 considering a concrete, practically relevant instance of a stream function in the realm of OBDA.

Chapter 5 introduces a general framework of stream processing in which streams are modelled as (possibly) infinite words over (possibly) infinite alphabets. Following the framework of [11], window functions are identified as the hidden structures for incrementally processing streams. I reformulate the main result of [11] as a representation result, develop a corresponding representation result for the case where the streams are excluded to be finite, and give further representation results when the windows have additional properties. A constructive, functional characterisation of bounded-memory window functions via recursion closes this chapter.

In Chap. 6 the focus is on high-level declarative stream processing within the STARQL query language framework (Streaming and Temporal ontology Access with a Reasoning-based Query Language). STARQL provides access to temporal and streaming data w.r.t. huge static data and an ontology—either with a very expressive ontology language or with light-weight description logics as required for sOBDA. After a short introduction to the syntax and semantics of STARQL, the chapter discusses the use of STARQL for sOBDA and provides an FOL rewritability result. Also it relates STARQL w.r.t. expressiveness to other ontology-based languages meant to provide access to temporal and streaming data.

Chapter 7, the last chapter before the overall conclusion, describes a representation result for reinterpretation operators, a special class of belief revision operators meant to be used for dissolving ambiguities. Additionally, I argue that the concept of reinterpretation (the hidden structure) is general enough to capture different forms of classical belief revision.

References

1. Alchourrón, C.E., Gärdenfors, P., Makinson, D.: On the logic of theory change: Partial meet contraction and revision functions. Journal of Symbolic Logic **50**, 510–530 (1985)
2. Areces, C.: Logic engineering. The case of description and hybrid logics. Ph.D. thesis, Institute for Logic, Language and Computation, University of Amsterdam, Amsterdam, The Netherlands (2000)

3. Arenas, M., Barceló Pablo, P., Libkin, L., Murlak, F.: Foundations of Data Exchange. Cambridge University Press, New York (2014)
4. van Benthem, J.: The Logic of Time: A Model-Theoretic Investigation into the Varieties of Temporal Ontology and Temporal Discourse, 2. edn. Reidel (1991)
5. Calvanese, D., De Giacomo, G., Lembo, D., Lenzerini, M., Poggi, A., Rodríguez-Muro, M., Rosati, R.: Ontologies and databases: The DL-Lite approach. In: Proceedings of the 5th International Reasoning Web Summer School (RW-09), *LNCS*, vol. 5689, pp. 255–356. Springer (2009)
6. ten Cate, B., Franconi, E., Seylan, I.: Beth definability in expressive description logics. Journal of Artificial Intelligence Research **48**(1), 347–414 (2013)
7. Delgrande, J.P., Schaub, T.: A consistency-based approach for belief change. Artificial Intelligence **151**(1–2), 1–41 (2003)
8. Fagin, R.: Generalized first-order spectra and polynomial-time recognizable sets. In: R. Karp (ed.) Computation and Complexity, *SIAM-AMS Proceedings*, vol. 7, pp. 43–73. American Mathematical Society (1974)
9. Fagin, R., Kolaitis, P.G., Miller, R.J., Popa, L.: Data exchange: Semantics and query answering. In: D. Calvanese, M. Lenzerini, R. Motwani (eds.) Proceedings of the 9th International Conference on Database Theory (ICDT-03), *LNCS*, vol. 2572, pp. 207–224. Springer (2003). DOI 10.1007/3-540-36285-1_14
10. Goldblatt, R.: Topoi, The Categorical Analysis of Logic, *Studies in Logic and the Foundations of Mathematics*, vol. 98. North-Holland, Amsterdam (1984)
11. Gurevich, Y., Leinders, D., Van Den Bussche, J.: A theory of stream queries. In: Proceedings of the 11th International Conference on Database Programming Languages, DBPL'07, pp. 153–168. Springer-Verlag, Berlin, Heidelberg (2007)
12. Halpern, J.Y., Harper, R., Immerman, N., Kolaitis, P.G., Vardi, M.Y., Vianu, V.: On the unusual effectiveness of logic in computer science. Bull. Symbolic Logic **7**(2), 213–236 (2001)
13. Immerman, N.: Descriptive complexity: A logician's approach to computer science. Notices of the American Mathematical Society **42**(10) (1995)
14. Immerman, N.: Descriptive Complexity. Graduate Texts in Computer Science. Springer-Verlag, New York (1999)
15. Jackson, D.: Software Abstractions: Logic, Language, and Analysis. The MIT Press, Cambridge, MA (2006)
16. Libkin, L.: Elements Of Finite Model Theory. Springer (2004)
17. Özçep, Ö.L.: Knowledge-base revision using implications as hypotheses. In: B. Glimm, A. Krüger (eds.) Proceedings of the 35th Annual German Conference on Artificial Intelligence (KI-12), LNCS, pp. 217–228. Springer Berlin Heidelberg (2012)
18. Özçep, Ö.L.: Bounded-memory stream processing. In: Proceedings of the 41st German AI Conference (KI 2018), Berlin (2018). Accepted for publication
19. Özçep, Ö.L., Grütter, R., Möller, R.: Nearness rules and scaled proximity. In: L.D. Raedt, C. Bessiere, D. Dubois (eds.) Proceedings of the 20th European Conference on Artificial Intelligence (ECAI-12), pp. 636–641 (2012)
20. Özçep, Ö.L., Möller, R.: Combining DL-Lite with spatial calculi for feasible geo-thematic query answering. In: Y. Kazakov, D. Lembo, F. Wolter (eds.) Proceedings of the 25th International Workshop on Description Logics (DL-12), vol. 846 (2012)
21. Özçep, Ö.L., Möller, R.: Computationally feasible query answering over spatio-thematic ontologies. In: Proceedings of the 4th International Conference on Advanced Geographic Information Systems, Applications, and Services (GEOProcessing-12) (2012)
22. Özçep, Ö.L., Möller, R.: Scalable geo-thematic query answering. In: P. Cudré-Mauroux, J. Heflin, E. Sirin, T. Tudorache, J. Euzenat, M. Hauswirth, J.X. Parreira, J. Hendler, G. Schreiber, A. Bernstein, E. Blomqvist (eds.) Proceedings of the 11th International Semantic Web Conference (ISWC-12), vol. 7649, pp. 658–673 (2012)
23. Özçep, Ö.L., Möller, R., Neuenstadt, C.: A stream-temporal query language for ontology based data access. In: Proceedings of the 37th German International Conference (KI-14), *LNCS*, vol. 8736, pp. 183–194. Springer International Publishing Switzerland (2014)

24. Özçep, Ö.L.: A representation theorem for spatial relations. In: B. Pfahringer, J. Renz (eds.) Proceedings of the 28th Australasian Joint Conference on Artificial Intelligence 2015 (AI-15), *LNAI*, vol. 9457, pp. 444–456 (2015)
25. Özçep, Ö.L.: Belief revision with bridging axioms. In: V. Rus, Z. Markov (eds.) Proceedings of the 30th International Florida Artificial Intelligence Research Society Conference (FLAIRS-17), Marco Island, pp. 104–109. AAAI Press (2017)
26. Özçep, Ö.L., Grütter, R., Möller, R.: Dynamics of a nearness relation–first results. In: Proceedings of the International Workshop on Spatio-Temporal Dynamics (SteDy-12) (2012)
27. Özçep, Ö.L., Möller, R.: Ontology based data access on temporal and streaming data. In: M. Koubarakis, G. Stamou, G. Stoilos, I. Horrocks, P. Kolaitis, G. Lausen, G. Weikum (eds.) Reasoning Web. Reasoning and the Web in the Big Data Era, *LNCS*, vol. 8714 (2014)
28. Özçep, Ö.L., Möller, R., Neuenstadt, C.: Stream-query compilation with ontologies. In: B. Pfahringer, J. Renz (eds.) Proceedings of the 28th Australasian Joint Conference on Artificial Intelligence (AI-15), *LNAI*, vol. 9457. Springer International Publishing (2015)
29. Randell, D.A., Cui, Z., Cohn, A.G.: A spatial logic based on regions and connection. In: Proceedings of the 3rd International Conference on Knowledge Representation and Reasoning (KR-92), pp. 165–176 (1992)
30. Russell, S.J., Norvig, P.: Artificial Intelligence – A Modern Approach. Prentice Hall (1995)
31. Satoh, K.: Nonmonotonic reasoning by minimal belief revision. In: Proceedings of the International Conference on Fifth Generation Computer Systems (FGCS-88), pp. 455–462. OHMSHA Ltd. Tokyo and Springer Verlag (1988)
32. Stone, M.H.: The theory of representations for boolean algebras. Transactions of the American Mathematical Society **40**(1), 37–111 (1936)
33. Uschold, M.: Where are the semantics in the semantic web? AI Magazine **24**(3), 25–36 (2003)
34. Vardi, M.Y.: From philosophical to industrial logics. In: Proceedings of the 3rd Indian Conference on Logic and Its Applications (ICLA-09), pp. 89–115. Springer-Verlag, Berlin, Heidelberg (2009). DOI 10.1007/978-3-540-92701-3_7
35. Völker, J., Vrandečić, D., Sure, Y., Hotho, A.: Aeon - an approach to the automatic evaluation of ontologies. Applied Ontology **3**(1-2), 41–62 (2008)
36. Wigner, E.: On the unreasonable effectiveness of mathematics in the natural sciences. Communications on Pure and Applied Mathematics **3**, 1–14 (1960)

Chapter 2
Preliminaries

Abstract This chapter provides necessary terminology and knowledge that is used throughout this monograph. In particular, it gives an overview of relevant bits of first-order logic, propositional logic, and description logics. In the context of the introduction to description logics, a technical overview of the paradigm of OBDA and the most prominent logic for OBDA, DL-Lite, is presented. Having this logical machinery, the notion of representation as used in this monograph is stated more formally. The preliminaries are concluded by a short introduction to the region connection calculus (RCC), which is considered in Chaps. 3 and 4.

2.1 Logics

Usually, four components have to be specified for a logic: syntax, semantics, model theoretical notions, and proof procedures/calculi. In this monograph, I deal mainly with the first three aspects with the focus on first-order logic (FOL) and two of its fragments, propositional logic as well the family of description logics.

2.1.1 First-Order Logic

First-order logic is a sufficiently expressive logic for modelling any kind of structure, be it mathematical structures or dynamic structures such as those that frequently appear in CS, e.g., transactional databases, processes, or streams. First-order logic is the reference logic for all other formal-specification languages and the other logics that are in the focus of this monograph.

Ö. L. Özçep, *Representation Theorems in Computer Science*,
https://doi.org/10.1007/978-3-030-25785-9_2

2.1.1.1 Syntax

An FOL *non-logical vocabulary* or *signature* σ consists of *(individual) constant symbols*, usually denoted by lower-case letters a, b, c, a_1, etc., *relation symbols*, usually denoted by uppercase symbols R, P etc., and *function symbols*, usually denoted by f, g etc. Each relation and function symbol has an associated arity—which is usually not explicitly mentioned because it is determined by the context in which the symbol is used. The non-logical symbols are intended to model arbitrary entities, and their semantics is fixed by structures as explicated below. In contrast to them, there are logical symbols with a fixed meaning. *Variables*, usually denoted by x, y etc., have a mixed status. As a convention, I refer to them as non-logical symbols. For each non-logical vocabulary two types of logical expression can be formulated, terms and formulae.

A *term* t is either a variable or a constant or it is of the form $f(t_1, \ldots, t_n)$, where $t_1, \ldots, t_n$ are terms, $n \in \mathbb{N}$, and f is an n-ary function symbol.

Formulae are either atomic or non-atomic. An *atomic formula* α is either of the form $\bot$ (representing a contradiction), $\top$ (representing a tautology), $t_i = t_j$ ("t_i is equal to t_j") or of the form $R(t_1, \ldots, t_n)$ where the t_i are terms and R is an n-ary relation symbol. If α and β are (atomic or non-atomic) formula, then the Boolean combinations are (non-atomic) formulae as well: $(\alpha \wedge \beta)$ ("α and β"); $(\alpha \vee \beta)$ ("α or β"); $(\alpha \rightarrow \beta)$ ("If α then β"); $(\alpha \leftrightarrow \beta)$ ("α iff β"). Also $\forall x\ \alpha$ ("For all x it holds that α") and $\exists x\ \alpha$ ("There is an x s.t. α") are (non-atomic) formula. Formulae that are either atomic or that are negations of an atomic formulae are called *literals*.

Quantifiers $\forall, \exists$ bind their associated variables. As usual one can define the occurrences of variables that are not bound, i.e., that are free in a formula. For example, in the formula $R(x, y) \wedge \forall x P(x)$ the (only) occurrence of y and the first occurrence of x is free, whereas the second occurrence of x is not free. A formula α with free occurrences of variables among $\{x_1, \ldots, x_n\}$ is denoted $\alpha(x_1, \ldots, x_n)$ or, even shorter, $\alpha(\mathbf{x})$ with the vector notation $\mathbf{x} = x_1, \ldots, x_n$.

Formulae of FOL can be used for defining queries (see the semantical definition below). One important subclass of FOL, which has been extensively investigated in database systems and in OBDA, are *conjunctive queries*, for short: CQs. These have the form $\exists y_1, \ldots, \exists y_m.\rho(y_1, \ldots, y_m, x_1, \ldots, x_n)$ consisting of a prefix of existential quantifiers $\exists$ and a conjunction ρ of atomic formula of the form $R_i(\mathbf{w}, \mathbf{z})$ where z_i, w_i stand either for variables contained in the set $\{y_1, \ldots, y_m, x_1, \ldots, x_n\}$ or for constants. The y_i are bound by the existential quantifiers, the free variables x_i are the output slots of the query. If the set of free variables is empty, the query is called a *Boolean query*.

A *union of CQs (UCQ)* is a disjunction of CQs α_i each with the same set of free variables. For a vector of constants $\mathbf{a}$ of length n and a formula α with free variables $\mathbf{x}$ of the same length, $\alpha(\mathbf{a})$ is the Boolean query resulting from substituting a_i for x_i, $i \in [n]$. Here and in the following I use the convenient abbreviation $[n] = \{1, \ldots, n\}$.

A formula without free variables is also called a *sentence*. The set of formulae for a given non-logical vocabulary σ is denoted *fml*(σ), the set of sentences is denoted *sent*(σ). An arbitrary set of sentences is sometimes called an *axiom set* or *set of axioms* or a *theory*. In the context of belief revision a finite set of sentences is sometimes called a *belief base* and sometimes *knowledge base*. If Ψ is a finite set of sentences $\alpha_1, \ldots, \alpha_n$, then $\bigwedge \Psi$ denotes the conjunction $(\ldots(\alpha_1 \wedge \alpha_2) \wedge \alpha_3) \wedge \ldots \alpha_n) \ldots)$ of all formulae in Ψ.

For any formula α, $symb(\alpha)$ is the set of non-logical symbols in α. For any set of sentences Ψ, $symb(\Psi) = \bigcup_{\alpha \in \Psi} symb(\alpha)$.

2.1.1.2 Semantics

The main semantical entity of FOL and other logics with a truth-conditional semantics is that of a *structure*. A σ-structure $\mathfrak{A}$ for a non-logical vocabulary σ is a pair $\mathfrak{A} = (A, \cdot^{\mathfrak{A}})$ where A is the domain, also denoted $dom(\mathfrak{A})$, and where $(\cdot)^{\mathfrak{A}}$ is a function, sometimes called the *denotation*, that maps every symbol $s \in \sigma$ to an entity of the correct type and arity: If s is a constant c, then $c^{\mathfrak{A}}$ is an element of A. If s is an n-ary relation symbols R, then $R^{\mathfrak{A}}$ is an n-ary relation $R^{\mathfrak{A}} \subseteq A^n = \underbrace{A \times \cdots \times A}_{n-\text{times}}$. If s is an n-ary function symbol f, then $f^{\mathfrak{A}}$ is an n-ary function $f^{\mathfrak{A}} \in A^{A^n} = \{g \mid g : A^n \longrightarrow A\}$. The set of all σ-structures is denoted $struct(\sigma)$.

Following the usual convention, I also sometimes describe a structure by explicitly enumerating the individuals, relations, and functions as in the following expression: $\mathfrak{A} = (A, {c_1}^{\mathfrak{A}}, \ldots, {c_m}^{\mathfrak{A}}, {R_1}^{\mathfrak{A}}, \ldots, {R_k}^{\mathfrak{A}}, {f_1}^{\mathfrak{A}}, \ldots, {f_l}^{\mathfrak{A}})$.

If a signature σ is made up of relation symbols only, then it is called a *relational* signature. In this case all σ-structures are also called relational. If the signature σ contains no relation symbols, i.e., it consists only of constants and function symbols, then it is called *algebraic*. In this case, a σ-structure is called an *algebra*.

Due to the ambivalent role of variables, the semantics of FOL requires adding also denotations for variables, that is functions that map a variable to an element of the domain. The denotation function for variables is also called an *assignment*. An FOL σ-*interpretation* is a pair $(\mathfrak{A}, \nu)$ of a σ-structure $\mathfrak{A}$ and an assignment ν which has as domain the set of variables mentioned in σ and which has as range the set $dom(\mathfrak{A})$. The set of all σ interpretations is denoted $Int(\sigma)$. The x-*variant* of an interpretation $\mathfrak{I}$ is denoted $\mathfrak{I}_{[x/d]}$. It is the same as $\mathfrak{I}$, except that $\mathfrak{I}_{[x/d]}$ maps x to the element $d \in dom(\mathfrak{A})$.

The semantics of terms is given recursively as follows: $\mathfrak{I}(c) = c^{\mathfrak{I}} = c^{\mathfrak{A}}$, $\mathfrak{I}(x) = x^{\mathfrak{I}} = \nu(x)$ and $\mathfrak{I}(f(t_1, \ldots, t_n)) = (f(t_1, \ldots, t_n))^{\mathfrak{I}} = f^{\mathfrak{A}}(\mathfrak{I}(t_1), \ldots, \mathfrak{I}(t_n))$.

Based on interpretations one can define recursively the *satisfaction relation* $\models$ between interpretations and formulae. For atoms the satisfaction relation is defined as follows: Not $\mathfrak{I} \models \bot$; $\mathfrak{I} \models \top$; $\mathfrak{I} \models t_1 = t_2$ iff $\mathfrak{I}(t_1) = \mathfrak{I}(t_2)$; and $\mathfrak{I} \models R(t_1, \ldots, t_n)$ iff $(\mathfrak{I}(t_1), \ldots, \mathfrak{I}(t_n)) \in R^{\mathfrak{A}}$. For non-atomic formulae the following recursive rules hold: $\mathfrak{I} \models \neg\alpha$ iff not $\mathfrak{I} \models \alpha$; $\mathfrak{I} \models (\alpha \wedge \beta)$ iff $\mathfrak{I} \models \alpha$ and $\mathfrak{I} \models \beta$; $\mathfrak{I} \models (\alpha \vee \beta)$ iff $\mathfrak{I} \models \alpha$ or $\mathfrak{I} \models \beta$; $\mathfrak{I} \models (\alpha \rightarrow \beta)$ iff: if $\mathfrak{I} \models \alpha$ then $\mathfrak{I} \models \beta$; $\mathfrak{I} \models (\alpha \leftrightarrow \beta)$ iff: $\mathfrak{I} \models \alpha$ iff $\mathfrak{I} \models \beta$; $\mathfrak{I} \models \forall x\, \alpha$ iff: for all $d \in A$: $\mathfrak{I}_{[x/d]} \models \alpha$; and $\mathfrak{I} \models \exists x\, \alpha$ iff: there is $d \in A$ such that $\mathfrak{I}_{[x/d]} \models \alpha$. If it is the case that $\mathfrak{I} \models \alpha$, then one of the following natural language wordings can be used: $\mathfrak{I}$ *fulfils* α, $\mathfrak{I}$ *satisfies* α, $\mathfrak{I}$ *models* α, $\mathfrak{I}$ *is a model of/for* α, $\mathfrak{I}$ *makes* α *true*. If not $\mathfrak{I} \models \alpha$, which is also written as $\mathfrak{I} \not\models \alpha$, then one can also say that $\mathfrak{I}$ *makes* α *false*. When α has no free variables, then for the specification of its models it is sufficient to consider only the underlying structure—ignoring the assignment. An interpretation is a *model* of a set of formula Ψ iff it is a model of each formula contained in Ψ. The set of models is denoted $[\![\Psi]\!]$.

A well known result is that the satisfaction relation for an interpretation with assignment ν only depends on the values for the open variables occurring in the formula. Hence, in order to determine whether an interpretation $\mathfrak{I}$ models a formula $\alpha(\mathbf{x})$ with open variables $\mathbf{x} = x_1, \ldots, x_n$, it is sufficient to consider partial assignments ν on $\mathbf{x}$. This motivates the alternative notation for satisfaction as $\mathfrak{A} \models \alpha[\mathbf{x}/\nu]$.

FOL formulae can be used to define *queries* over a signature σ. In a very general sense, queries Q are just functions of the form

$$Q : struct(\sigma) \longrightarrow struct(\tau)$$

Here I define the notion of a query induced by an FOL formula. For this let $\alpha(\mathbf{x})$ be an FOL formula with open variables $\mathbf{x} = x_1, \ldots, x_n$. Let $\tau = (\{ans\})$ be the special target signature τ consisting of the n-ary *answer* predicate *ans* and consider the set of structures $struct_H(\tau)$ over τ where the domain of each structure consists of exactly the constants σ_{ind} of the source signature σ. Then, the *FOL query induced by the FOL formula* $\alpha(\mathbf{x})$ is defined as the function

$$\begin{aligned} Q_{\alpha(\mathbf{x})} &: struct(\sigma) \longrightarrow struct_H(\tau) \\ &\qquad \mathfrak{A} \mapsto (\ \sigma_{const}\ ,\ \{(a_1, \ldots, a_n) \in (\sigma_{ind})^n \mid \mathfrak{A} \models \alpha(\mathbf{a})\}\ ) \end{aligned}$$

A more suggestive and widely used equivalent notation is that of the *set of answers* $\alpha(\mathbf{x})$ *of a formula over a structure* $\mathfrak{A}$ defined as follows:

$$ans(\alpha(\mathbf{x}), \mathfrak{A}) = \{\mathbf{a} = (a_1, \ldots, a_n) \in (\sigma_{ind})^n \mid \mathfrak{A} \models \alpha(\mathbf{a})\}$$

The reason why I consider not bindings of the variables $\mathbf{x}$ over the tuples of elements from the domains of the source structures but bindings over the tuples of individual constants over σ_{ind} is that I will consider query answering mainly in the context of OBDA where the approach with constants is more convenient (see below).[1]

I will follow the usual practice of referring also to the formulae that define a query by the term "query"—as already done above when discussing conjunctive queries. To distinguish both readings I sometimes talk about "query formulae" and of "query functions".

[1] A very general notion of an FOL query, where next to an arbitrary target signature τ the use of products of σ structures is allowed (similar to the approach of interpretability in model theory) can be found in [9, p. 19])

2.1.1.3 Model Theoretical Notions

A formula is *valid* if it is true for all interpretations. It is *contradictory* if it is false for all interpretations. For arbitrary sets of sentences Ψ and sentences α one says that Ψ *entails* α or that α *follows from* Ψ, for short $\Psi \models \alpha$, iff every model of Ψ is a model of α, i.e., iff $[\![\Psi]\!] \subseteq [\![\alpha]\!]$.

Two sets of sentences Ψ_1 and Ψ_2 are said to be equivalent iff they have the same models. In this case one uses the short notation $\Psi_1 \equiv \Psi_2$.

The entailment relation induces a *consequence operator* $Cn(\cdot)$ for sets of sentences Ψ defined by $Cn(\Psi) = \{\alpha \in sent(\sigma) \mid \Psi \models \alpha\}$. Sometimes one considers only consequences w.r.t. a sub-signature σ': $Cn^{\sigma'}(\Psi) = \{\alpha \in sent(\sigma') \mid \Psi \models \alpha\} = Cn(\Psi) \cap sent(\sigma')$. A consequence operator according to Tarski fulfils for any set (of formulae) X, Y the following properties: 1. $X \subseteq Cn(X)$ (reflexivity) 2. If $X \subseteq Y$, then $Cn(X) \subseteq Cn(Y)$ (monotonicity) and 3. $Cn(Cn(X)) = Cn(X)$ (idempotence).

Relations between structures can be captured by various notions. The ones that are used in this monograph are given below.

A simple relation between two σ structures $\mathfrak{A}$, $\mathfrak{B}$ is that of one being the *substructure* of the other: $\mathfrak{A}$ is a substructure of $\mathfrak{B}$ iff $dom(\mathfrak{A}) \subseteq dom(\mathfrak{B})$, for all constants $c \in \sigma$ $c^{\mathfrak{A}} = c^{\mathfrak{B}}$, for all function symbols f and all elements $d_1, \ldots, d_n \in dom(\mathfrak{A})$: $f^{\mathfrak{A}}(d_1, \ldots, d_n) \in dom(\mathfrak{A})$ and $f^{\mathfrak{A}} = f^{\mathfrak{B}} \upharpoonright (dom(\mathfrak{A}))^n$ and for all relation symbols R: $R^{\mathfrak{A}} = R^{\mathfrak{B}} \cap (dom(\mathfrak{A}))^n$.

Two structures $\mathfrak{A}$, $\mathfrak{B}$ over the same signature σ are said to be *isomorphic* iff there is a bijective function $\pi : dom(\mathfrak{A}) \rightarrow dom(\mathfrak{B})$ such that the following holds:

1. For all constants c: $\pi(c^{\mathfrak{A}}) = c^{\mathfrak{B}}$
   (c is denoted by the corresponding objects)
2. For all n-ary function symbols R and n-tuples $(a_1, \ldots, a_n)$: $\pi(f^{\mathfrak{A}}(a_1, \ldots, a_n) = (f^{\mathfrak{B}}(\pi(a_1), \ldots, \pi(a_n))$.
3. For all n-ary predicate symbols R and n-tuples $(a_1, \ldots, a_n)$: $R^{\mathfrak{A}}(a_1, \ldots, a_n)$ iff $R^{\mathfrak{B}}(\pi(a_1), \ldots, \pi(a_n))$

If two structures $\mathfrak{A}$, $\mathfrak{B}$ are isomorphic then this is denoted by $\mathfrak{A} \simeq \mathfrak{B}$. In the case of algebras only the first two conditions are relevant.

A weaker notion of structure preservation is that of a homomorphism. A function $h : dom(\mathfrak{A}) \longrightarrow dom(\mathfrak{B})$ between two structures $\mathfrak{A}$, $\mathfrak{B}$ over the same signature σ is called a *homomorphism* iff all of the following conditions hold:

- for any constant c: $h(c^{\mathfrak{A}}) = c^{\mathfrak{B}}$
- for any n-ary relation R and elements $a_1, \ldots, a_n$ from $dom(\mathfrak{A})$: If $R^{\mathfrak{A}}(a_1, \ldots, a_n)$, then also $R^{\mathfrak{B}}(h(a_1), \ldots, h(a_n))$
- for any n-ary function symbol f and elements $a_1, \ldots, a_n$ from $dom(\mathfrak{A})$: $f^{\mathfrak{A}}(a_1, \ldots, a_n) = f^{\mathfrak{B}}(h(a_1), \ldots, h(a_n))$.

In this case the short notation $h : \mathfrak{A} \stackrel{hom}{\longrightarrow} \mathfrak{B}$ is used.

An important proposition, that is relevant for strict OBDA, states that UCQs are preserved under homomorphisms.

Proposition 2.1 *Let* $h : \mathfrak{A} \overset{hom}{\longrightarrow} \mathfrak{B}$ *be a homomorphism and* Q *be a UCQ. Then: for all tuples* $\mathbf{a}$ *from the domain of* $\mathfrak{A}$*: if* $\mathbf{a} \in Q(\mathfrak{A})$*, then* $h(\mathbf{a}) \in Q(\mathfrak{B})$

A function $h : dom(\mathfrak{A}) \longrightarrow dom(\mathfrak{B})$ between two structures $\mathfrak{A}, \mathfrak{B}$ over the same signature σ is called a *strong homomorphism* iff it is a homomorphism and additionally the following conditions hold:

- for any $b_1, \ldots, b_n \in dom(\mathfrak{B})$ with $R^{\mathfrak{B}}(b_1, \ldots, b_n)$ there are $a_1, \ldots, a_n \in dom(\mathfrak{A})$ with $h(a_i) = b_i$ (for all $i \in \{1, \ldots, n\}$) and $R^{\mathfrak{A}}(a_1, \ldots, a_n)$.
- for any $b_1, \ldots, b_n, b_{n+1} \in dom(\mathfrak{B})$ with $f^{\mathfrak{B}}(b_1, \ldots, b_n) = b_{n+1}$ there are $a_1, \ldots, a_n, a_{n+1} \in dom(\mathfrak{A})$ with $h(a_i) = b_i$ (for all $i \in \{1, \ldots, n+1\}$) and $f^{\mathfrak{A}}(a_1, \ldots, a_n) = a_{n+1}$.

2.1.2 Propositional Logic

If an FOL signature σ is chosen to contain only relation symbols of arity 0 (but no constants and function symbols) and if further the logical vocabulary is restricted to the Boolean connectors $\neg, \vee, \wedge, \rightarrow, \leftrightarrow$, the outcome is a *propositional-logic signature*. On the basis of such a signature the syntax and semantics of propositional logic is exactly the semantics and syntax for FOL presented in the previous section. For the convenience of the reader, I redefine the syntax and semantics, using the simplifications that result from the restriction to a propositional logic signature.

2.1.2.1 Syntax

The only syntactical category in propositional logic is that of a sentence. The 0-ary relational symbols are also called *propositional symbols*. A signature consisting only of propositional symbols will be denoted by $\mathcal{P}$ instead of σ. Furthermore, instead of the convention applied for FOL, propositional symbols will be denoted with lowercase letters such as p, q, r etc. The set of sentences $sent(\mathcal{P})$ is already defined by the FOL rules for formulae that do not mention the quantifier. That is, the set of sentences is given by the following context free grammar:

$$\alpha ::= p \mid \neg\alpha \mid (\alpha \wedge \alpha) \mid (\alpha \vee \alpha) \mid (\alpha \rightarrow \alpha) \mid (\alpha \leftrightarrow \alpha) \mid \bot \mid \top$$

A *clause* is a disjunction of literals. Sometimes clauses are written in set-wise manner, e.g., $p_1 \vee \neg p_2 \vee p_3$ is also represented as the set $\{p_1, \neg p_2, p_3\}$. With respect to this representation a clause is a *subclause* of another clause iff it is a subset of this clause. A conjunction of literals is called a *dual clause*. A propositional formula is in *conjunctive normal form (CNF)* iff it is a conjunction of clauses. A disjunction of conjunctions of literals is a formula in *disjunctive normal form (DNF)*. Given a formula $\alpha \in sent(\mathcal{P})$ and a subset $S \subseteq \mathcal{P}$ of symbols, the *clausal closure* of α w.r.t. S is the set $clause^S(\alpha)$ of clauses that have only symbols from S and that follow from α.

2.1.2.2 Semantics

As propositional variables are 0-ary relation symbols, there is only one possible interpretation for them: either the empty relation $\emptyset$ or the singleton set $\{()\}$ containing the 0-ary tuple (). In this context, usually, the empty set is denoted by the Boolean value 0 (standing for false) and the singleton with the 0-ary tuple by the Boolean value 1 (standing for true). In propositional logic one does not have to specify the domain of a structure nor an assignment for variables (as there are no variables.) So the notion of an FOL interpretation in the case of proposition logic is a function that assigns truth values 0, 1 to propositional symbols in $\mathcal{P}$. Sticking to the terminology introduced for FOL, the set of interpretations over the propositional variables $\mathcal{P}$ is denoted $Int(\mathcal{P})$. Instead of $\mathfrak{I} \models \beta$, in propositional logic I prefer the alternative specification with Boolean truth values: $\mathfrak{I}(\alpha) = 1$ iff $\mathfrak{I} \models \alpha$ and $\mathfrak{I}(\alpha) = 0$ iff $\mathfrak{I} \not\models \alpha$. $\mathfrak{I}_{[p/v]}$ for $v \in \{0, 1\}$ denotes the variant of $\mathfrak{I}$ that assigns to the propositional variable p the value v.

In the chapter on belief revision (Chap. 7), an alternative representation of propositional-logic interpretations is going to be used: interpretations will be identified with the set of propositional symbols which are assigned the value 1. For example, let $\mathfrak{I}$ be an interpretation over $\mathcal{P} = \{p, q, r\}$ with $\mathfrak{I}(p) = 1$, $\mathfrak{I}(q) = 0$, $\mathfrak{I}(r) = 1$, then $\mathfrak{I}$ is identified with the set $\{p, r\}$.

The truth value of a formula α depends only on the symbols occurring in it. Hence, in order to determine the truth value of α it is sufficient to consider $Int(S)$ instead of $Int(\mathcal{P})$, where $S \subseteq \mathcal{P}$ is a set of symbols with $symb(\alpha) \subseteq S$.

2.1.2.3 Consequences Relative to a Symbol Set

Sometimes one is interested in axiomatising all the consequences of a formula w.r.t. a given set of propositional variables. This is the case, e.g., for the reinterpretation operators considered in this monograph. I define two operators Θ_S and Θ'_S that, given a formula α and a set S of symbols $S \subseteq \mathcal{P}$, compute a formula axiomatising all consequences of α that do not contain symbols in S. For Θ'_S, the argument α has to be transformed in DNF while Θ_S does not presuppose such a transformation. These operators will be used as technical aids for calculating belief-revision results based on reinterpretation (see Chap. 7).

Let α be a formula, $dnf(\alpha)$ a formula equivalent to α represented as a set of clauses and $S \subseteq \mathcal{P}$. Furthermore, I assume that $dnf(\alpha)$ is a reduced formula in the sense that it does not contain a contradictory dual clause. $\Theta'_S(\alpha)$ results from $dnf(\alpha)$ by substituting all literals over S in $dnf(\alpha)$ by the logical constant $\top$: or equivalently: delete all literals in $dnf(\alpha)$ that contain a symbol of S. The empty dual clause is interpreted as $\top$.

Θ_S is based on substituting symbols in S by truth value assignments. Let $\mathfrak{I} \in Int(S)$ be given, then the formula $\alpha_{\mathfrak{I}}$ is defined as follows: substitute all occurrences of $p \in S$ in α where $p^{\mathfrak{I}} = \mathfrak{I}(p) = 1$ by $\top$, else $\bot$ is substituted for p. For example, let $\alpha = (p \wedge q) \vee (r \wedge s)$ and $S = \{p, r\}$ and $\mathfrak{I} \in Int(S)$ with $\mathfrak{I} : p \mapsto 1$, $\mathfrak{I} : r \mapsto 0$, then $\alpha_{\mathfrak{I}} = (\top \wedge q) \vee (\bot \wedge s)$. Now Θ_S is defined as follows: Let $S \subseteq symb(\alpha)$. Then $\Theta_S : \alpha \mapsto \bigvee_{\mathfrak{I} \in Int(S)} \alpha_{\mathfrak{I}}$. For arbitrary $S \subseteq \mathcal{P}$ let $\Theta_S(\alpha) = \Theta_{symb(\alpha) \cap S}(\alpha)$. The following facts concerning Θ'_S and Θ_S can be easily proved.

Proposition 2.2 $\alpha \models \Theta'_S(\alpha)$ *and* $\alpha \models \Theta_S(\alpha)$

Proof See p. 39. □

Proposition 2.3 *Let* $S \subseteq \mathcal{P}$. *For all formulae* α *over* $\mathcal{P}$ *and* $\theta_S \in \{\Theta'_S, \Theta_S\}$*:* $Cn^{\mathcal{P} \setminus S}(\alpha) = Cn^{\mathcal{P} \setminus S}(\theta_S(\alpha))$

Proof See p. 39. □

As a corollary to Proposition 2.2 and 2.3 the logical equivalence of $\Theta'_S(\alpha)$ and $\Theta_S(\alpha)$ follows.

Corollary 2.4 $\Theta'_S(\alpha) \equiv \Theta_S(\alpha)$.

Proof As $\alpha \models \Theta'_S(\alpha)$ and $Cn^{P \setminus S}(\alpha) = Cn^{\mathcal{P} \setminus S}(\Theta_S(\alpha))$, $\Theta_S(\alpha) \models \Theta'_S(\alpha)$. Similarly $\alpha \models \Theta_S(\alpha)$ and $Cn^{\mathcal{P} \setminus S}(\alpha) = Cn^{\mathcal{P} \setminus S}(\Theta'_S(\alpha))$ entail the fact that $\Theta'_S(\alpha) \models \Theta_S(\alpha)$. So, $\Theta'_S(\alpha) \equiv \Theta_S(\alpha)$. □

2.1.3 Description Logics

Description Logics (DLs) are logics for use in knowledge representation with special attention on a good balance of expressibility and feasibility of reasoning services. The expressivity of most DLs lies between that of FOL, which is expressive but for which important problems such as validity are not even decidable, and that of propositional logic, which is moderately expressive but for which most relevant algorithmic problems are decidable.

DLs are mainly used as representation means for *ontologies* which are the backbone of the paradigm of ontology-based data access (OBDA) as well as of the semantic web architecture. Due to this role, the syntax of description logics is meant to provided convenient means to model concept descriptions. Formally, concept descriptions correspond to FOL formulae that have one open variable and that are tree shaped with this open variable as root.

Restricting a FOL signature to constant symbols (also called *individual constants*), unary relation symbols (called *atomic concepts* or *concept symbols*) or *concept names*, and binary symbols (called *atomic roles* or *role symbols* or *role names*) results in a *DL signature*. Based on a DL signature, four categories of DL expressions can be defined, concept descriptions *concepts*(σ), role descriptions *roles*(σ), abox axioms *abox-axioms*(σ), and tbox axioms *tbox-axioms*(σ). Depending on the kinds of reasoning services in which a DL is used and depending on the required expressivity/feasibility relation, different DLs can be defined. The various DLs differ regarding the allowed set of concept and role constructors, the (non-)use of concrete domains and data types, and on the set of constraints for building axioms from concepts, roles, and constants.

In DL speak, an *ontology* (in the wider sense) is a triple $O = \langle \sigma, \mathcal{A}, \mathcal{T} \rangle$ with a DL signature σ, a finite set of abox axioms $\mathcal{A} \subseteq \mathit{abox\text{-}axioms}(\sigma)$ (set of assertional axioms), and a finite set $\mathcal{T} \subseteq \mathit{tbox\text{-}axioms}(\sigma)$ of tbox axioms (set of terminological axioms). When the signature is clear from the context, then also $\mathcal{T} \cup \mathcal{A}$ is called the ontology. Note that some authors use the term "ontology" only for the tbox $\mathcal{T}$.

The semantics of DLs is based on structures as defined for FOL, the only difference being that denotations for a DL signature are specified, where constants are denoted by individuals of a domain, concept symbols are denoted by subsets of the domain, and roles are denoted by binary relations over the domain.

In this monograph, I deal mainly with a family of DLs that is a family of lightweight DLs tailored towards strict OBDA, namely the DL-Lite family [4]. In the following I describe the syntax and semantics for one member of this family, the logic DL-Lite$^{\sqcap}_{\mathcal{F},\mathcal{R}}$.

Definition 2.5 (DL-Lite$^{\sqcap}_{\mathcal{F},\mathcal{R}}$) Let σ be a DL signature $\sigma = \sigma_{RN} \cup \sigma_{CN} \cup \sigma_{ind}$ where σ_{RN} is a set of role symbols and $P \in \sigma_{RN}$, σ_{CN} is a set of concept symbols and $A \in \sigma_{CN}$, and where σ_{ind} is a set of individual constants and $a, b \in \sigma_{ind}$.
Syntax.

roles(σ):	$R \longrightarrow P \mid P^-$
concepts(σ):	$B \longrightarrow A \mid \exists R \quad C_l \longrightarrow B \mid C_l \sqcap B \quad C_r \longrightarrow B \mid \neg B$
tbox-axioms(σ):	$C_l \sqsubseteq C_r$, (funct R), $R_1 \sqsubseteq R_2$
abox-axioms(σ):	$A(a)$, $R(a, b)$
Constraint:	If R occurs in a functionality axiom, then R and R^- do not occur as R_2 in a role inclusion axiom $R_1 \sqsubseteq R_2$.

Semantics.
Let $\mathfrak{I}$ be a σ-interpretation $\mathfrak{I} = (\Delta^{\mathfrak{I}}, \cdot^{\mathfrak{I}})$, where the denotation function $\cdot^{\mathfrak{I}}$ specifies $A^{\mathfrak{I}} \subseteq \Delta^{\mathfrak{I}}$ for all $A \in \sigma_{CN}$, $c^{\mathfrak{I}} \in \Delta^{\mathfrak{I}}$ for all $c \in \sigma_{ind}$ and $c_1^{\mathfrak{I}} = c_2^{\mathfrak{I}}$ iff $c_1 = c_2$ (unique name assumption), and $R^{\mathfrak{I}} \subseteq \Delta^{\mathfrak{I}} \times \Delta^{\mathfrak{I}}$ for all $R \in \sigma_{RN}$.
The semantics of concepts descriptions w.r.t. $\mathfrak{I}$ is as follows: $(C \sqcap D)^{\mathfrak{I}} = C^{\mathfrak{I}} \cap D^{\mathfrak{I}}$; $\neg B = \Delta^{\mathfrak{I}} \setminus B^{\mathfrak{I}}$; and $(\exists R)^{\mathfrak{I}} = \{d \in \Delta^{\mathfrak{I}} \mid \text{ there is } e \in \Delta^{\mathfrak{I}} \text{ s.t. } (d, e) \in r^{\mathfrak{I}}\}$. The semantics for tbox axioms is as follows: $\mathfrak{I} \models C \sqsubseteq D$ iff $C^{\mathfrak{I}} \subseteq D^{\mathfrak{I}}$; $\mathfrak{I} \models R_1 \sqsubseteq R_2$ iff $R_1^{\mathfrak{I}} \subseteq R_2^{\mathfrak{I}}$; and $\mathfrak{I} \models$ (funct R) iff $R^{\mathfrak{I}}$ is functional in its first argument. The semantics for abox axioms is as follows: $\mathfrak{I} \models A(b)$ iff $a^{\mathfrak{I}} \in A^{\mathfrak{I}}$; and $\mathfrak{I} \models R(a, b)$ iff $(a^{\mathfrak{I}}, b^{\mathfrak{I}}) \in R^{\mathfrak{I}}$.

2.2 Ontology-Based Data Access

One of the most important standard reasoning services considered in OBDA are satisfiability checking of an ontology and query answering. Satisfiability checking means checking whether there is an interpretation modelling an ontology $\mathcal{T} \cup \mathcal{A}$.

The usual semantics for query answering w.r.t. an ontology is based on the entailment relation. Given an FOL query $\alpha(\mathbf{x})$ with open variables $\mathbf{x} = x_1, \ldots, x_n$ the set of *certain answers w.r.t. an ontology* is the set of all bindings for $\mathbf{x}$ such that the formula $\alpha[\mathbf{x}/\mathbf{a}]$ follows from the ontology:

$$cert(\alpha(\mathbf{x}), \mathcal{A} \cup \mathcal{T}) = cert(\alpha(\mathbf{x}), (\sigma, \mathcal{A}, \mathcal{T})) = \{\mathbf{a} \in (\sigma_{ind})^n \mid \mathcal{T} \cup \mathcal{A} \models \alpha[\mathbf{x}/\mathbf{a}]\}$$

If α is a Boolean query, then the only outcomes are $\{()\}$ which is interpreted as truth value 1 or $\emptyset$ which is interpreted as the truth value 0.

In strict OBDA, the aim is to reduce query answering over an ontology to model-checking on the abox. This is realised by rewriting the given query w.r.t. the tbox to a new query that is evaluated over the associated minimal model of an abox. The same holds for satisfiability checking: checking the satisfiability of the union of the tbox and abox is reduced to model checking a (Boolean) query over the minimal model of the abox.

The associated minimal model abox $\mathcal{A}$ mentioned above is constructed as the minimal Herbrand model and is denoted $DB(\mathcal{A})$. More concretely, the minimal Herbrand model $DB(\mathcal{A}) = (\Delta, \cdot^{\mathfrak{I}})$ for an abox $\mathcal{A}$ is defined as follows:

- Δ = set of constants occurring in $\mathcal{A}$;
- $c^{\mathfrak{I}} = c$ for all constants;
- $A^{\mathfrak{I}} = \{c \mid A(c) \in \mathcal{A}\}$;
- $r^{\mathfrak{I}} = \{(c, d) \mid R(c, d) \in \mathcal{A}\}$

I define the notion of FOL rewritability in a quite general way, making use of different languages. A further generalisation, which also covers rewritability over temporal and streaming domains, is going to be discussed in Chap. 5.

Definition 2.6 Let QL_1 and QL_2 be query languages over the same signature and OL be an ontology language. QL_1 allows for QL_2-rewriting of query answering w.r.t. OL iff for all queries α in QL_1 and tboxes $\mathcal{T}$ in OL there exists a query $\alpha_{\mathcal{T}}$ in QL_2 such that for all aboxes $\mathcal{A}$ it holds that:

$$cert(\alpha, \mathcal{A} \cup \mathcal{T}) = ans(\alpha_{\mathcal{T}}, DB(\mathcal{A}))$$

A particularly interesting case is QL_2 = first-order logic (FOL) queries. A well-known fact [4] is: UCQs are FOL rewritable w.r.t. DL-Lite ontologies.

Actually, FOL rewriting in this sense can be divided into two steps. In the first step only the elimination of the tbox $\mathcal{T}$ is required, i.e., certain answering w.r.t. the union of tbox and abox is reduced to certain answering over the abox only. The second step then requires reducing certain answering w.r.t. the abox to answering over the associated minimal DB of the abox. Formally, the first step is expressed by the following equation:

$$cert(\alpha, \mathcal{T} \cup \mathcal{A}) = cert(\alpha_{\mathcal{T}}, \mathcal{A})$$

If there is no other constraint on the target query language than the constraint that it consists of an FOL formula, then this form of rewriting can be accomplished always: rewrite the tbox $\mathcal{T}$ equivalently into a finite FOL formula β and consider $\alpha_{\mathcal{T}} = \beta \rightarrow \alpha$. Because I do not have further constraints on the target query language, I consider in the following only rewriting to the associated minimal model of an abox.

The proofs for FOL rewritability use the idea of constructing a model that is universal. A well-known construction principle for building universal models is the *chase* construction, which is extensively used in OBDA but originally goes back to data base theory and is also extensively used in Data Exchange [7, 1].

Definition 2.7 A model of a theory is called a *universal model* iff it can be embedded homomorphically into any other model of the theory.

This property of universal models makes them interesting for answering UCQs. Together with the preservation property according to Proposition 2.1 it follows that universal models capture all answers to a UCQ that are contained in all sets of answers of the query for each model of the theory. Actually this means that universal models capture the certain answers of a UCQ w.r.t. a theory (an ontology).

Proposition 2.8 *For any universal model* $\mathfrak{I} \models \mathcal{T} \cup \mathcal{A}$ *(if it exists) and any UCQ* q *it holds that:*

$$ans(q, \mathfrak{I}) = cert(q, \mathcal{T} \cup \mathcal{A})$$

The lightweight description logic DL-Lite$^{\sqcap}_{\mathcal{F},\mathcal{R}}$ defined above is tailored towards FOL rewritability as verified by the following theorem.

Theorem 2.9 (Mainly [4, Thm 4.14, Thm 5.15]) *DL-Lite*$^{\sqcap}_{\mathcal{F},\mathcal{R}}$ *is FOL rewritable w.r.t. satisfiability as well as w.r.t. answering UCQs.*

In the following, the chase construction for DL-Lite$^{\sqcap}_{\mathcal{F},\mathcal{R}}$ is described in more detail as it will be used in an adapted fashion for the proof of FOL rewritability for a spatial extension of DL-Lite$^{\sqcap}_{\mathcal{F},\mathcal{R}}$ introduced in Chap. 4. The idea of the chase construction is to repair the abox with respect to the constraints formulated in the tbox. If, e.g., the tbox contains the axiom $A_1 \sqsubseteq A_2$ and the abox contains $A_1(a)$ but not $A_2(a)$, then it is enriched by the atom $A_2(a)$. This procedure is applied stepwise to yield a sequence of aboxes S_i starting with the original abox $\mathcal{A}$ as S_0. The resulting set of abox axioms $\bigcup S_i$ may be infinite but induces a canonical model $can(\mathcal{O})$ for the abox and the tbox axioms being used in the chasing process. I sketch the chase construction for DL-Lite$^{\sqcap}_{\mathcal{F},\mathcal{R}}$.

Let $\mathcal{T}$ be a DL-Lite tbox, let $\mathcal{T}_p$ be the subset of positive inclusion (PI) axioms in $\mathcal{T}$, i.e, those GCIs that contain no negation and let $\mathcal{A}$ be an abox and $\mathcal{O} = \mathcal{T} \cup \mathcal{A}$. Chasing will be carried out with respect to PIs only. Let $S_0 = \mathcal{A}$. Let S_i be the set of abox axioms constructed so far and α be a PI axiom in $\mathcal{T}_p$. Let α be of the form $A_1 \sqsubseteq A_2$ and let $\beta \in S_i$ (resp. $\beta \subseteq S_i$) be an abox axiom (resp. set of abox axioms). The PI axiom α is called applicable to β if β is of the form $A_1(a)$ and $A_2(a)$ is not in S_i. The applicability of other PI axioms of the form $B \sqsubseteq C$ is defined similarly [4, Def. 4.1, p. 287]. If the left-hand side of the PI is a conjunction of base concepts, e.g., if the PI is of the form $A_1 \sqcap \cdots \sqcap A_n \sqsubseteq A_0$, and if β is $\{A_1(a), \ldots, A_n(a)\}$ and $A_0(a)$ is not in S_i, then PI is applicable to β.

As there may be many possible applications of PI axioms to atoms and sets of atoms, one has to impose an order on the tbox axioms and the (finite) subsets of the abox. So I assume that all strings over the signature σ of the ontology O and some countably infinite set of new constants C_{ch} are well ordered. Such a well ordering exists and has the order type of the natural numbers $\mathbb{N}$. This ordering is different from the one of [4], but it can also be used also for infinite aboxes and it can handle concept conjunction. If there is a PI axiom α applicable to an atom β in S_i, one takes the minimal pair (α, β) with respect to the ordering and produces the next level $S_{i+1} = S_i \cup \{\beta_{new}\}$. Here β_{new} is the atom that results from applying the chase rule for (α, β) as listed in Def. 2.10. The primed constants (in particular the a' in Def. 2.10) are the chasing constants from C_{ch}.

Definition 2.10 (Chasing rules for DL-Lite$^{\sqcap}_{\mathcal{F},\mathcal{R}}$)

If $\alpha = A_1 \sqsubseteq A_2$ and $\beta = A_1(a)$ then $\beta_{new} = A_2(a)$
If $\alpha = A_1 \sqsubseteq \exists R$ and $\beta = A_1(a)$ then $\beta_{new} = R(a, a')$
If $\alpha = \exists R \sqsubseteq A$ and $\beta = R(a, b)$ then $\beta_{new} = A(a)$
If $\alpha = \exists R_1 \sqsubseteq \exists R_2$ and $\beta = R_1(a, b)$ then $\beta_{new} = R_2(a, a')$
If $\alpha = R_1 \sqsubseteq R_2$ and $\beta = R_1(a, b)$ then $\beta_{new} = R_2(a, b)$
If $\alpha = A_1 \sqcap \cdots \sqcap A_n \sqsubseteq A_0$ and $\beta = \{A_1(a), \ldots, A_n(a)\}$ then $\beta_{new} = A_2(a)$
(and similarly for other PIs of the form $B_1 \sqcap \cdots \sqcap B_n \sqsubseteq C$)

The chase is defined by $chase(O) = chase(\mathcal{T}_p \cup \mathcal{A}) = \bigcup_{i \in \mathbb{N}} S_i$. The canonical model $can(O)$ is the minimal Herbrand model of $chase(O)$. The canonical model $can(O)$ is a universal model of $\mathcal{T}_p \cup \mathcal{A}$ with respect to homomorphisms. In particular this entails that answering a UCQ w.r.t. $\mathcal{T}_p \cup \mathcal{A}$ can be reduced to answering $Q^{can(O)}$ w.r.t. $DB(\mathcal{A})$. More concretely, (some finite closure $cln(\mathcal{T})$ of) the negative inclusions axioms and the functionality axioms are only relevant for checking the satisfiability of $cln(\mathcal{T})$. The details for the construction of $cln(\mathcal{T})$ can be found in the appendix on p. 39.

A convenient method for proving non-FOL rewritability uses well-known facts from descriptive complexity [9], which investigates correspondences between computational resource required to solve problems and the logics to represent the problems. Usually, in descriptive complexity one is interested in the problem of answering queries over finite structures (which correspond to databases). In many practical cases the size of the query is much smaller than the size of the structure, hence one defines the notion of *data complexity* which measures (time and space) resources only w.r.t. the size of the structure but not the size of the query.

The relevant complexity class for FOL rewritability is AC^0. Intuitively, this class consists of problems solvable in parallel constant time on polynomially many processors. Formally, AC^0 is defined in the computation model of Boolean circuits: it is the set of problems solvable by families of circuits with constant depth, polynomial size and using NOT gates, unlimited-fanin AND gates and OR gates.

Now, the descriptive complexity correspondence most relevant for OBDA can be stated as follows:

Proposition 2.11 *The data complexity for FOL query answering is AC^0. In the other direction: any query problem in AC^0 can be encoded as answering an FOL query.*

Together, the two assertions of the proposition state that FOL *captures* AC^0.

The following relations between AC^0 and other more well-known complexity classes hold:

$$AC^0 \subsetneq TC^0 \subseteq \text{LOGSPACE} \subseteq \text{NLOGSPACE} \subseteq \text{PTIME} \subseteq \{NP, coNP\} \subseteq PH \subseteq PSPACE \ldots$$

In this monograph I will mainly use the facts that $AC^0 \subsetneq \text{LOGSPACE} \subseteq NP$.

Strict OBDA is intended to be used for accessing data stored in a database, called X in the following paragraph. Therefore, to apply the idea of rewriting, the data in the database have to be represented as an abox. This is accomplished by a set of mappings $\mathcal{M}$ consisting of rules which determine how rows in a table are translated to abox axioms. The details of the mappings do not matter for this monograph. Assume that the abox defined on X and for the mappings $\mathcal{M}$ is denoted $\mathcal{A}(\mathcal{M}, X)$. The abox for which query answering is conducted is just this $\mathcal{A}(\mathcal{M}, X)$. But this abox is actually not materialised, i.e., not really generated in strict OBDA. Rather, answering a query α on $\mathcal{A}(\mathcal{M}, X)$ requires another rewriting step termed *unfolding*: given the mappings $\mathcal{M}$, the query α is unfolded to a DB query which can be executed on X. Now, the point is that the query language of the database (usually SQL) has some properties that lead to constraints for performant unfolding. This is indeed the case for most relational database and stream management systems that are equipped with SQL like query languages. These have the property of *domain independence*: a query α is *domain independent* iff for all interpretations $\mathfrak{I}, \mathfrak{J}$ such that $\mathfrak{I}$ is a substructure of $\mathfrak{J}$: $ans(\alpha, \mathfrak{I}) = ans(\alpha, \mathfrak{J})$ [2].

2.3 Notion of Representation

The notion of representation used in this paragraph is oriented at the notion of representation as used in the well-known result of Stone for Boolean algebras [15]. Hence, before giving my definition of representation I recapitulate the theorem of Stone.

A Boolean algebra is defined as a special algebra, i.e., a structure which consists of a domain A and constants as well as functions over A. As usual, 0-ary functions can be identified with constant elements of A.

Definition 2.12 (Boolean Algebra) A structure $\mathfrak{A} = (A^{\mathfrak{A}}, +^{\mathfrak{A}}, \cdot^{\mathfrak{A}}, -^{\mathfrak{A}}, 0^{\mathfrak{A}}, 1^{\mathfrak{A}}) = (A, +, \cdot, -, 0, 1)$ with two binary functions $+, \cdot$, a unary function $-$ and two constants $0, 1$ is called a *Boolean algebra* iff it fulfils the following constraints:

$$
\begin{array}{ll}
x + y = y + x & x \cdot y = y \cdot x \\
x + (y + z) = (x + y) + z & x \cdot (y \cdot z) = (x \cdot y) \cdot z \\
x + 0 = x & x \cdot 1 = x \\
x + (-x) = 1 & x \cdot (-x) = 0 \\
x + (y \cdot z) = (x + y) \cdot (x + z) & x \cdot (y + z) = (x \cdot y) + (x \cdot z)
\end{array}
$$

Set algebras are special Boolean algebras of the form $\mathfrak{A} = (pow(A), \cup, \cap, \overline{}, \emptyset, A)$ where the domain of the structure consists of the powerset of a given set A, $+$ is instantiated by the union operator, $\cdot$ is instantiated by the intersection operator, $\overline{}$ is the complement of a set, 0 stands for the empty set $\emptyset$, and 1 stands for the whole set A.

Theorem 2.13 *Every Boolean algebra is isomorphic to a set algebra.*

The theorem says that all set algebras are indeed Boolean algebras (which can be verified easily using naive set theory) and that for any Boolean algebra one can construct a set algebra that has the same structure. Hence, set algebras represent the whole set of Boolean algebras, they can be called canonical because the set-theoretical algebraic functions are general enough to capture any Boolean algebra. The proof of the theorem is a nice exercise in universal algebra (or model theory) which constructs semantical objects (ultrafilters) considering maximality conditions on syntactical objects (the axioms defining Boolean algebras).

I give a (semi-)formal definition of a representation theorem on which all the results of this monograph are based on.

Definition 2.14 Let σ be a signature and $Rns \subseteq Rnd \subseteq struct(\sigma)$ be sets of σ-structures, *Rns* standing for the representing structures and *Rnd* for the structures to be represented. For any σ-structure $\mathfrak{A} \in Rnd$ and $\mathfrak{B} \in Rns$ let $maps(\mathfrak{A}, \mathfrak{B}) = dom(\mathfrak{B})^{dom(\mathfrak{A})}$ be the set of all possible functions from the domain of $\mathfrak{A}$ to the domain of $\mathfrak{B}$ and let $mapsAll(X, Y) := \bigcup_{\mathfrak{A} \in X, \mathfrak{B} \in Y} maps(\mathfrak{A}, \mathfrak{B})$. A *representation theorem* then has the following components:

1. a simple construction principle for *Rns*,
2. a function $F : dom(Rnd) \longrightarrow mapsAll(Rns, Rnd)$ such that for all $\mathfrak{B} \in Rnd$: $F(\mathfrak{B}) \in maps(\mathfrak{A}, \mathfrak{B})$ for some $\mathfrak{A} \in Rns$,
3. and a structure-preservation property that any function $F(\mathfrak{B})$ for all $\mathfrak{B} \in Rnd$ must fulfil.

The semi-formality of the definition comes from the non-formalised notions of a "simple construction principle" and that of a structure-preserving property.

Within the course of the monograph the instantiations for the application-specific construction principles and the structure-preserving properties will become clear. Here, I illustrate the definition in case of Stone's representation theorem:

1. $\sigma = \{+, \cdot, -, 0, 1\}$ = signature of Boolean algebras
2. Rnd = Boolean algebras
3. Rns = set algebras

4. *Rns* constructed with power sets
5. function F given by ultrafilters of maximally consistent sets
6. structure preserving property: isomorphism.

As mentioned in the introduction, also FOL rewritability theorems are related to representation theorems. FOL rewritability of a query amounts to finding for any $\mathcal{T}$ and query α an FOL query $\alpha_{\mathcal{T}}$ such that $cert(\alpha, \mathcal{T} \cup \mathcal{A}) = ans(\alpha_{\mathcal{T}}, DB(\mathcal{A}))$ for all aboxes $\mathcal{A}$. The main tool for showing the existence of $\alpha_{\mathcal{T}}$ is to show the existence of a universal model, using, e.g., the chase construction. The query $\alpha_{\mathcal{T}}$ captures exactly the information stored in the universal model which represents the common positive information of all models of the theory.

So the components for a representation theorem in the context of FOL rewritability are the following:

1. σ = a DL signature of an ontology $(\sigma, \mathcal{T}, \mathcal{A})$
2. $Rnd = [\![\mathcal{T} \cup \mathcal{A}]\!]$
3. $Rns = \{\Im_u\}$ with an arbitrary universal model $\Im_u$ of $\mathcal{T} \cup \mathcal{A}$
4. function F giving for any model $\Im \in Rnd$ a homomorphism from $\Im_u$ into it.
5. structure preserving property: homomorphism.

As the reader might have foreseen, it is possible to define other relevant representation notions by changing the type of mapping between the representing structures and the represented structures. I mention here only the notion of a *prime model* (see any text book on model theory such as [5]) where the mapping is that of an embedding, i.e. an injective strong homomorphism.

2.4 The Region Connection Calculus (RCC)

The region connection calculus (RCC) is a calculus for qualitative spatial reasoning with regions. Actually, behind this term there is a whole family of calculi RCCi with different levels of expressiveness. The basic theory according to [12] assumes a primitive binary relation $\mathsf{C}(x, y)$, read as "x connects with y", which in turn is defined on regions and not on points. The relation $\mathsf{C}(x, y)$ is reflexive and symmetric.

$$\forall x.\mathsf{C}(x, x) \text{ and } \forall x, y[\mathsf{C}(x, y) \rightarrow \mathsf{C}(y, x)]$$

The family of calculi RCCi (for $i \in \{1, 2, 3, 5, 8\}$) are characterized by sets $\mathcal{B}_{RCCi}$ of i *base relations* $\mathcal{B}_{RCCi} = \{r_1, \ldots, r_i\}$ which are defined on the base of $\mathsf{C}(x, y)$ and which have the JEPD-property: they are jointly exhaustive and pairwise disjoint. Formally this means that for all x, y the fact $r_i(x, y)$ holds exactly for one relation r_i. The most expressive calculus RCC8 is based upon the set of base relations $\mathcal{B}_{RCC8}$ defined as follows:

$$\mathcal{B}_{RCC8} = \{\mathsf{DC}, \mathsf{EC}, \mathsf{PO}, \mathsf{EQ}, \mathsf{TPP}, \mathsf{NTPP}, \mathsf{TPPi}, \mathsf{NTPPi}\}$$

The definitions of the base relations of $\mathcal{B}_{RCC8}$ and other relations that I use in this monograph are given as predicate logical sentences in Fig. 2.1.

$$\begin{aligned}
\mathsf{DC}(x, y) &\leftrightarrow \neg\mathsf{C}(x, y) && \text{(disconnected from)}\\
\mathsf{P}(x, y) &\leftrightarrow \forall z(\mathsf{C}(z, x) \rightarrow \mathsf{C}(z, y)) && \text{(part of)}\\
\mathsf{PP}(x, y) &\leftrightarrow \mathsf{P}(x, y) \wedge \neg\mathsf{P}(y, x) && \text{(proper part of)}\\
\mathsf{EQ}(x, y) &\leftrightarrow \mathsf{P}(x, y) \wedge \mathsf{P}(y, x) && \text{(equal to)}\\
\mathsf{O}(x, y) &\leftrightarrow \exists z(\mathsf{P}(z, x) \wedge \mathsf{P}(z, y)) && \text{(overlaps)}\\
\mathsf{PO}(x, y) &\leftrightarrow \mathsf{O}(x, y) \wedge \neg\mathsf{P}(x, y) \wedge \neg\mathsf{P}(y, x) && \text{(partially overlaps)}\\
\mathsf{DR}(x, y) &\leftrightarrow \neg\mathsf{O}(x, y) && \text{(discrete from)}\\
\mathsf{EC}(x, y) &\leftrightarrow \mathsf{C}(x, y) \wedge \neg\mathsf{O}(x, y) && \text{(externally connected to)}\\
\mathsf{TPP}(x, y) &\leftrightarrow \mathsf{PP}(x, y) \wedge \exists z(\mathsf{EC}(z, x) \wedge \mathsf{EC}(z, y)) && \text{(tangential proper part of)}\\
\mathsf{NTPP}(x, y) &\leftrightarrow \mathsf{PP}(x, y) \wedge \neg\exists z(\mathsf{EC}(z, x) \wedge \mathsf{EC}(z, y)) && \text{(nontangential proper part of)}\\
\mathsf{Pi}(x, y) &\leftrightarrow \mathsf{P}(y, x) && \text{(inverse of P)}\\
\mathsf{PPi}(x, y) &\leftrightarrow \mathsf{PP}(y, x) && \text{(inverse of PP)}\\
\mathsf{TPPi}(x, y) &\leftrightarrow \mathsf{TPP}(y, x) && \text{(inverse of TPP)}\\
\mathsf{NTPPi}(x, y) &\leftrightarrow \mathsf{NTPP}(y, x) && \text{(inverse of NTPP)}
\end{aligned}$$

Fig. 2.1 Definitions of RCC relations according to [12, p. 167]

For the convenience of the reader Fig. 2.2 illustrates the base relations.

Note that for two regions x, y, $\mathsf{EQ}(x, y)$ means that x, y cover the same area in space. RCC8 allows for models in which $\mathsf{EQ}(x, y)$ may hold even if $x \neq y$, that is even if x, y denote different objects. These models are called *non-strict* models in [14]. This distinction is useful for a fine-grained representation of administrative regions (see Sect. 3.8). For example, a municipality is not just a spatially extended object but some "abstract" entity with specific legal obligations, functions etc.

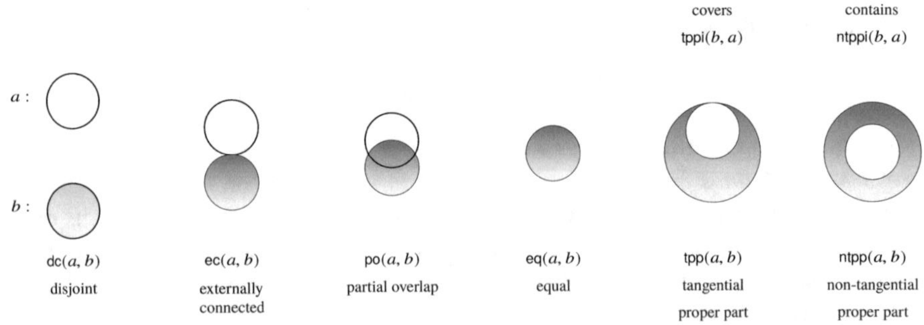

Fig. 2.2 RCC8 Base relations

Beside the definitions of base relations the axiom system of Randell and colleagues [12] contains the axiom of non-atomicity. It states that every region has a non-tangential proper part—which immediately leads to an infinite set of regions.

$$\forall x \exists y.\mathsf{NTPP}(y, x)$$

Randell and colleagues [12] also define binary functions for regions. One of these is the sum function for regions x, y which results in the union z of x, y. It is defined by the following axiom:

$$\forall x, y, z[\mathsf{sum}(x, y) = z \leftrightarrow \forall w(\mathsf{C}(w, z) \leftrightarrow (\mathsf{C}(w, x) \vee \mathsf{C}(w, y)))]$$

That means, z is the sum of x and y if and only if the following holds: any region w connects with the sum if and only if it connects with one of the summands. Instead of $\mathsf{sum}(x, y)$ I also use the set theoretic notation $x \cup y$ and assume that the sum function is extended to any finite number of arguments (using the associativity of sum) so that also $\bigcup_{i \in I} x_i$ for any finite index set I is defined. The other Boolean functions are complement, intersection, and difference. Complementation is defined by:

$$\forall x, y.\ \mathsf{compl}(x) = y \leftrightarrow \forall z[(\mathsf{C}(z, y) \leftrightarrow \neg\mathsf{NTPP}(z, x)) \wedge (\mathsf{O}(z, y) \leftrightarrow \neg\mathsf{P}(z, x))]$$

The intersection or product $\mathsf{prod}(x, y)$ (also denoted set theoretically by $x \cap y$) is defined by the following axiom:

$$\mathsf{prod}(x, y) = z \leftrightarrow \forall u[\mathsf{C}(u, z) \leftrightarrow \exists v(\mathsf{P}(v, x) \wedge \mathsf{P}(v, y) \wedge \mathsf{C}(u, v))]$$

And the difference $\mathsf{diff}(x, y)$ (also denoted set theoretically by $x \setminus y$) is defined as follows:

$$\mathsf{diff}(x, y) = w \leftrightarrow \forall z[\mathsf{C}(z, w) \leftrightarrow \mathsf{C}(z, \mathsf{prod}(x, \mathsf{compl}(y)))]$$

I call the set consisting exactly of the axioms in this section *the axiom set for the Boolean region connection calculus* and denote it by Ax_{BRCC}. I decided to work with the axiom set Ax_{BRCC} and not a specific RCC model upon because I wanted to keep the model as general as possible and make the deduction of nearness properties in the propositions comprehensible by referring to the axioms (see Chap. 4). Nonetheless, the reader may think of a specific model of $Ax_{\mathcal{BRCC}}$ when following the proofs for Sect. 3.8, for example, the model that results from interpreting regions as regular closed subsets of $\mathbb{R}^2$ equipped with the usual topology. In this model, regular closed sets x, y are connected iff they share a common point, i.e., $\mathsf{C}(x, y)$ iff $x \cap y \neq \emptyset$. Intuitively, the regularity restriction means that the regions do not have cuts or pointed holes.

For qualitative spatial reasoning with RCC, [6] introduced so-called *composition tables*. The composition $\circ$ of two relations r_1 and r_2 is defined as follows:

$$r_1 \circ r_2 = \{(x, y) \mid \exists z.r_1(x, z) \wedge r_2(z, y)\}$$

Given relation facts $r_1(x, y)$ and $r_2(y, z)$ with relations r_1, r_2 from a set of relations $\{r_1, r_2, \ldots, r_i\}$, a composition table enables one to look up the possible relations that can hold between x and z. So with the composition of two base relations, in most cases, only indefinite knowledge of spatial configurations follows. The spatial configuration $r_1(x, z) \vee \cdots \vee r_n(x, z)$ for base relations r_j in $\mathcal{B}_{RCCi}$ is also written as $\{r_1, \ldots, r_n\}(x, z)$, and the set $\{r_1, \ldots, r_n\}$ is called a general RCCi relation. Let Rel_{RCCi} be the set of all 2^i general RCCi relations. An RCCi *(constraint) network* consists of assertions of the form $\{r_1, \ldots, r_n\}(x, y)$.

As an example of a composition table entry in RCC8, which is relevant for the engineering bureau scenario in Chap. 4, I mention the table entry for the pair (tpp, tppi): tpp; tppi = {dc, ec, po, tpp, tppi, eq} which is described in Ax_{RCC8} by $\forall x, y, z.\mathsf{tpp}(x, y) \wedge \mathsf{tppi}(y, z) \rightarrow \{\mathsf{dc, ec, po, tpp, tppi, eq}\}(x, z)$.

For the convenience of the reader, I show in Table 2.1 the composition table for RCC8. The set of 8 base relations in RCC8 is denoted by $\mathcal{B}_{RCC8}$. Note that any pair (r_1, r_2) with entry $r_3^1, \ldots, r_3^k$ corresponds to an FOL sentence of the form:

$$\forall x, y, z[(r_1(x, y) \wedge r_2(y, z)) \rightarrow r_3^1(x, z) \vee \cdots \vee r_3^k(x, z)]$$

∘	DC	EC	PO	TPP	NTPP	TPPi	NTPPi	EQ
DC	$\mathcal{B}_{RCC8}$	DR, PO, PP	DR, PO, PP	DR, PO, PP	DR, PO, PP	DC	DC	DC
EC	DR, PO, PPi	DR, PO, TPP, TPi	DR, PO, PP	EC, PO, PP	PO, PP	DR	DC	EC
PO	DR, PO, PPi	DR, PO, PPi	$\mathcal{B}_{RCC8}$	PO, PP	PO, PP	DR, PO, PPi	DR, PO, PPi	PO
TPP	DC	DR	DR, PO, PP	PP	NTPP	DR, PO, TPP, TPi	DR, PO, PPi	TPP
NTPP	DC	DC	DR, PO, PP	NTPP	NTPP	DR, PO, PP	$\mathcal{B}_{RCC8}$	NTPP
TPPi	DR, PO, PPi	EC, PO, PPi	PO, PPi	PO, TPP, TPi	PO, PP	PPi	NTPPi	TPPi
NTPPi	DR, PO, PPi	PO, PPi	PO, PPi	PO, PPi	O	NTPPi	NTPPi	NTPPi
EQ	DC	EC	PO	TPP	NTPP	TPPi	NTPPi	EQ

Table 2.1 Composition table for RCC8 (TPi stands for {TPPi, EQ})

I also mention here the composition table for the low resolution logics RCC2 and RCC3 as I am going to refer to them in later chapters. Their base relations are given by the sets $\mathcal{B}_{RCC3}$ = {DR, EQ, ONE} and $\mathcal{B}_{RCC2}$ = {DR, O}, and their weak compositions are defined as shown in Fig. 2.3. The discreteness relation DR is the same as {DC, EC}, the overlapping-but-not-equal relation ONE is equal to {PO, NTPP, TPP, NTPPi, TPPi} and the overlapping relation O is given by {ONE, EQ}. Note that in the definitions of the base relations (of RCC3 and RCC2) I followed the author of [16] and not [8]. But the composition tables for both definitions are identical.

In Chap. 4 I am not going to refer to the connect relation C nor to the summation operator. Rather I am going to consider simpler axiomatizations of the RCC calculus.

Definition 2.15 (Axiom system schema Ax_{RCCi}) For all $i \in \{2, 3, 5, 8\}$ the axiom set Ax_{RCCi} contains the following axioms:

$$\{\forall x, y. \bigvee_{r \in \mathcal{B}_{RCCi}} r(x, y)\} \cup \qquad \text{(joint exhaustivity)}$$
$$\{\forall x, y. \bigwedge_{r_1, r_2 \in \mathcal{B}_{RCCi}, r_1 \neq r_2} r_1(x, y) \rightarrow \neg r_2(x, y)\} \cup \qquad \text{(pairwise disjointness)}$$
$$\{\forall x, y, z. r_1(x, y) \wedge r_2(y, z) \rightarrow r_3^1(x, z) \vee \cdots \vee r_3^k(x, z) \mid r_1 ; r_2 = \{r_3^1, \ldots, r_3^k\}\} \qquad \text{(weak composition axioms)}$$

For $i \in \{3, 5, 8\}$ additionally the axiom $\forall x \mathsf{EQ}(x, x)$ (reflexivity of EQ) is contained. For $i = 2$ the axiom $\forall x \mathsf{O}(x, x)$ (reflexivity of O) is contained.

In particular, the axioms state the JEPD-property of the base relations (each pair of regions x, y is related over exactly one base relation) and describe the (weak) composition of two base relations (denoted by ;) according to the composition table for RCCi.

A practically relevant question for which composition tables are used is whether a network is satisfiable with respect to the RCC8-axioms. For example, the network $\{\mathsf{tpp}(a, b), \mathsf{tpp}(b, c), \mathsf{tpp}(a, c)\}$ is satisfiable whereas $\{\mathsf{tpp}(a, b), \mathsf{ntpp}(b, c), \mathsf{tpp}(a, c)\}$ is not satisfiable. Testing the satisfiability of networks can be carried out by path consistency algorithms [10].

By a translation into the modal logic S4 it can be shown that the satisfiability test for RCC8-networks is in NPTIME [3]. By showing that the decidability problem 3SAT—i.e., the problem of deciding whether a propositional formula in CNF with clauses that contain at most 3 literals has a model (alias: is satisfiable)—is reducible to the satisfiability of RCC8-networks, the NPTIME-hardness follows [13]. Consequently, testing the satisfiability of arbitrary RCC8-networks is NPTIME-complete and therefore a computationally intensive task. Tractability of the satisfiability of RCC8-networks can be gained by restricting the labels to a specific subclass of all RCC8-relations Rel_{RCC8}. A maximally tractable subset of RCC8-relations in the sense that the satisfiability test is in PTIME is defined in [13]. If one constrains the RCC8-networks to so-called conjunctive RCC8-networks [8], i.e. networks that contain only a base relation from $\mathcal{B}_{RCC8}$ or the whole set $\mathcal{B}_{RCC8}$ as label, then the complexity of the satisfiability test can be more specifically described as lying in NC [11]. Intuitively, NC (Nick's Class) is the class of problems that are decidable in polylogarithmic time on a parallel computer with a polynomial number of pro-

;	DR	O
DR	$\mathcal{B}_{RCC2}$	$\mathcal{B}_{RCC2}$
O	$\mathcal{B}_{RCC2}$	$\mathcal{B}_{RCC2}$

;	DR	ONE	EQ
DR	$\mathcal{B}_{RCC3}$	{DR, ONE}	DR
ONE	{DR, ONE}	$\mathcal{B}_{RCC3}$	ONE
EQ	DR	ONE	EQ

Fig. 2.3 Composition tables for RCC2 and RCC3

cessors.[2] This can be made precise by Boolean circuit complexity. NC is the class of all problems that can be decided by a uniform system of Boolean circuits with a polylogarithmic depth and polynomial size (polynomial number of gates.)

References

1. Arenas, M., Barceló Pablo, P., Libkin, L., Murlak, F.: Foundations of Data Exchange. Cambridge University Press, New York (2014)
2. Avron, A.: Constructibility and decidability versus domain independence and absoluteness. Theoretical Computer Science **394**(3), 144–158 (2008)
3. Bennett, B.: Modal logics for qualitative spatial reasoning. Logic Journal of the IGPL **4**(1), 23–45 (1996)
4. Calvanese, D., De Giacomo, G., Lembo, D., Lenzerini, M., Poggi, A., Rodríguez-Muro, M., Rosati, R.: Ontologies and databases: The DL-Lite approach. In: Proceedings of the 5th International Reasoning Web Summer School (RW-09), *LNCS*, vol. 5689, pp. 255–356. Springer (2009)
5. Chang, C., Keisler, H.: Model Theory. Studies in Logic and the Foundations of Mathematics. Elsevier Science (1990)
6. Cohn, A.G., Bennett, B., Gooday, J., Gotts, N.M.: Qualitative Spatial Representation and Reasoning with the Region Connection Calculus. Geoinformatica **1**, 275–316 (1997)
7. Fagin, R., Kolaitis, P.G., Miller, R.J., Popa, L.: Data exchange: Semantics and query answering. In: D. Calvanese, M. Lenzerini, R. Motwani (eds.) Proceedings of the 9th International Conference on Database Theory (ICDT-03), *LNCS*, vol. 2572, pp. 207–224. Springer (2003). DOI 10.1007/3-540-36285-1_14
8. Grigni, M., Papadias, D., Papadimitriou, C.H.: Topological inference. In: Proceedings of the 14th International Joint Conference on Artificial Intelligence (IJCAI-95), pp. 901–907 (1995)
9. Immerman, N.: Descriptive Complexity. Graduate Texts in Computer Science. Springer-Verlag, New York (1999)
10. Mackworth, A.K.: Consistency in networks of relations. Artificial Intelligence pp. 99–118 (1977)
11. Nebel, B.: Computational properties of qualitative spatial reasoning: First results. In: Proceedings of the 19th Annual German Conference on Artificial Intelligence (KI-95), KI '95, pp. 233–244. Springer-Verlag, London, UK, UK (1995)
12. Randell, D.A., Cui, Z., Cohn, A.G.: A spatial logic based on regions and connection. In: Proceedings of the 3rd International Conference on Knowledge Representation and Reasoning (KR-92), pp. 165–176 (1992)
13. Renz, J., Nebel, B.: On the complexity of qualitative spatial reasoning: A maximal tractable fragment of the region connection calculus. Artificial Intelligence **108**, 69–123 (1999). DOI 10.1016/S0004-3702(99)00002-8
14. Stell, J.G.: Boolean connection algebras: A new approach to the region-connection calculus. Artificial Intelligence **122**(1-2), 111–136 (2000). DOI http://dx.doi.org/10.1016/S0004-3702(00)00045-X
15. Stone, M.H.: The theory of representations for boolean algebras. Transactions of the American Mathematical Society **40**(1), 37–111 (1936)
16. Wessel, M.: On spatial reasoning with description logics - position paper. In: S.T. Ian Horrocks (ed.) Proceedings of the International Workshop on Description Logics (DL-02), *CEUR Workshop Proceedings*, vol. 53 (2002). Http://CEUR-WS.org/Vol-53/

[2] Though it is known that $NC \subseteq P$, it is not known whether $NC \subsetneq P$.

Appendix

Construction of $cln(\mathcal{T})$

I give details on the construction of $cln(\mathcal{T})$. The finite closure $cln(\mathcal{T})$ of the negative inclusions axioms and the functionality axioms are (only) relevant for checking the satisfiability of the ontology which can be tested by a simple FOL query. With induction on the stepwise construction of the chase one can show that $can(O)$ is a model of the whole ontology O iff the negative inclusion axioms and functionality axioms are in accordance with the original abox. The authors of [4] define the *negative closure* $cln(\mathcal{T})$ in order to capture all possible conflicts [4, Def. 4.7, p. 292]. I extend their definition of the negative closure to the logic DL-Lite$^{\sqcap}_{\mathcal{F},\mathcal{R}}$ and reformulate the definition in an alternative representation that fits better to the extensions $cl_{\perp}$ that we will use in the proofs of some propositions. Let $\bar{B}$ be an abbreviation for $B_1 \sqcap \cdots \sqcap B_n$.

1. All functionality axioms of $\mathcal{T}$ are in $cln(\mathcal{T})$.
2. For all negative inclusions $\bar{B} \sqsubseteq \neg B \in \mathcal{T}$ let $\bar{B} \sqcap B \sqsubseteq \perp \in cln(\mathcal{T})$.
3. If $\bar{B} \sqsubseteq B \in \mathcal{T}$ and $B \sqcap B' \sqsubseteq \perp \in cln(\mathcal{T})$, then $\bar{B} \sqcap B' \sqsubseteq \perp \in cln(\mathcal{T})$.
4. If $P_1 \sqsubseteq P_2 \in \mathcal{T}$ and $\exists P_2 \sqcap \bar{B} \sqsubseteq \perp \in cln(\mathcal{T})$, then $\exists P_1 \sqcap \bar{B} \sqsubseteq \perp \in cln(\mathcal{T})$.
5. If $P_1 \sqsubseteq P_2 \in \mathcal{T}$ and $\exists P_2^- \sqcap \bar{B} \sqsubseteq \perp \in cln(\mathcal{T})$, then $\exists P_1^- \sqcap \bar{B} \sqsubseteq \perp \in cln(\mathcal{T})$.
6. Let $X := \{\exists P \sqsubseteq \neg \exists P, \exists P^- \sqsubseteq \neg \exists P^-\}$.
   If $X \cap cln(\mathcal{T}) \neq \emptyset$, then $X \subseteq cln(\mathcal{T})$.

The FOL query for testing the satisfiability is built as the disjunction of boolean queries q_τ for every $\tau \in cln(\mathcal{T})$ in the following way: if $\tau = (\text{funct}\, R)$ then $q_\tau = \exists x, y, z R(x, y) \wedge R(x, z) \wedge y \neq z$; if $\tau = A \sqsubseteq \perp$, then $q_\tau = \exists x. A(x)$. For the other $\tau \in cln(\mathcal{T})$ the query q_τ is defined similarly [4, p. 296].

Proof of Proposition 2.2

Let $\alpha^{\Im} = 1$. Then $(dnf(\alpha))^{\Im} = 1$. Hence there is a dual clause kl in $dnf(\alpha)$ such that $kl^{\Im} = 1$. By definition there is a subclause $kl' \subseteq kl$ in $\Theta'_S(\alpha)$. Therefore $(\Theta'_S(\alpha))^{\Im} = 1$. If $\alpha^{\Im} = 1$, then $(\alpha_{\Im})^{\Im} = 1$ and hence $(\Theta_S(\alpha))^{\Im} = 1$.

Proof of Proposition 2.3

Proof of "$\supseteq$": This is a consequence of Proposition 2.2.

Proof of "$\subseteq$": Let $\theta = \Theta_S$. Let $\beta \notin Cn^{P \setminus S}(\theta(\alpha))$. Then there is an assignment $\Im_1$ for $\bigvee_{\Im \in Int(S)} \alpha_v$ such that $\Im_1 \models \bigvee_{\Im \in Int(S)} \alpha_{\Im}$ and $\Im_1 \models \neg\beta$. The first relation entails that there is a $\Im \in Int(S)$ such that $\Im_1 \models \alpha_{\Im}$. Define a new interpretation $\Im_2$ with $p^{\Im_2} = p^{\Im}$ for all $p \in S$ and $p^{\Im_2} = p^{\Im_1}$ for all other symbols. Then $\Im_2 \models \neg\beta$ and $\Im_2 \models \alpha$, and hence $\beta \notin Cn^{\mathcal{P} \setminus S}(\alpha)$.

Now let $\theta = \Theta'_S$. We have to show: for all δ with $symb(\delta) \subseteq \mathcal{P} \setminus S$ it is the case that if $\alpha \models \delta$, then also $\theta(\alpha) \models \delta$. This assertion is equivalent to the assertion that for δ with $symb(\delta) \subseteq \mathcal{P} \setminus S$ it is the case that if $\theta(\alpha) \cup \{\neg\delta\}$ has a model, $\alpha \cup \{\neg\delta\}$ has a model, too. Let $\mathfrak{I}$ be a model of $\theta(\alpha) \cup \{\neg\delta\}$. That means that there exists a dual clause kl' in $\theta(\alpha)$ with $(kl')^{\mathfrak{I}} = 1$. In $dnf(\alpha)$ there is a dual clause kl such that substituting all literals of S in kl by $\top$ results in kl'. Let $\mathfrak{I}'$ be a modification of $\mathfrak{I}$ defined by: for all $p \notin S$ let $p^{\mathfrak{I}'} = p^{\mathfrak{I}}$. For all $p \in S$ let $p^{\mathfrak{I}'} = 1$ if $p \in kl$ and $p^{\mathfrak{I}'} = 0$ if $\neg p \in kl$. (We may assume that not at the same time $p, \neg p \in kl$.) Then $kl^{\mathfrak{I}'} = 1$ and hence $\alpha^{\mathfrak{I}'} = 1$ follows. Because $\mathfrak{I}$ changes at most symbols in S, it is also the case that $(\neg\delta)^{\mathfrak{I}'} = 1$.

Chapter 3
Representing Spatial Relatedness

Abstract Spatial relations have been investigated in various related areas such as qualitative spatial reasoning, geographic information science, general topology, and many more. Most of the results are specific constructions of spatial relations that fulfil some required properties. Representation results setting up axioms that capture exactly the desired properties of the qualitative spatial relations are rare. This chapter provides a representation theorem for a particular binary relation of spatial relatedness which is motivated by the observation that human perception of nearness depends on a hierarchy of levels or contexts.

3.1 Introduction

Spatial relations have been investigated in various inter-related areas such as qualitative spatial reasoning [27], geographic information science [28], general topology [18], and others. Most of the results achieved are specific constructions of spatial relations that fulfil some desired properties—which may vary according to the application/modelling context. Although the axiomatic method is a well-proven approach for the description of entities, results on setting up axioms that capture the desired properties of spatial relations are rare. And results that characterise spatial relations in the sense that they state a representation theorem are missing.

Following the general representation methodology of this monograph, this chapter describes how to fill this gap with a semantic analysis of a specific class of spatial relations: not only does it set up an axiom set that the intended spatial relation should fulfil but it takes a deeper look into the structure of the models for the axioms. This is the general idea underlying proofs for representation theorems, namely that of systematically characterising the models by grouping them into disjoint, mathematically well-defined classes. A particularly interesting case is the one in which the set of models is described by exactly one class of models built according to some construction principle. In this case, the axioms really characterise the intended concepts, providing a canonical representation according to the construction principle

Ö. L. Özçep, *Representation Theorems in Computer Science*,
https://doi.org/10.1007/978-3-030-25785-9_3

of the class. As stated in the introduction of this monograph, a nice side-product of a representation theorem is that unintended models, which could result from an incomplete axiomatisation, are excluded.

The spatial relations for which this monograph gives a representation theorem are defined on the basis of a special structure, a hierarchical structure of nested partitions [12, 19, 21]. Typical examples of such total orders of nested partitions are made up of administrative units where the administrative units in a rougher granularity (e.g., districts) are the unions of administrative units of the lower level (e.g., municipalities). For example, think of two partitions of Switzerland, where the first partition consists of municipalities and where the second consists of districts. All districts are municipalities or are unions of two or more municipalities.

The interest in such types of spatial relations stems from observations regarding the context-dependency of spatial relatedness: the criterion for deciding whether a is considered spatially related to b depends on the type of the object that has b as its spatial extension. If b is a natural object such as a mountain, then the spatial criteria (be it geometric, topological, or metric) for identifying objects as spatially related or not may depend on scaling contexts of (big) natural borders such as those of forests or rivers etc. But if the object with spatial extension b is a non-natural artifact such as a house, then different criteria have to be taken into account: say borders made up by cadastral data.

The situation, as the house example demonstrates, may even be more complicated: it may be the case that the same spatial area b (house area) is the spatial extension of two different objects: the house considered as pure geometrical object or the house considered as a legal object which has to adhere to planning laws. Depending on which objects are relevant for the use case different criteria are relevant in order to decide whether an object is spatially related to a house: in the first case, a purely metric criterion is in order, in the second case the district or even the country in which the house is situated is in order.

In this chapter, only objects of the same type are considered. With respect to the example above this means that either houses are considered as pure geometrical objects only or that houses are considered as legal objects only. Hence, for spatial relatedness there is one context criterion fixed according to which two objects are considered to be related or not. This criterion is formalised by nested partitions of a spatial domain X. A partition provides a granularity or scale w.r.t. which spatial relatedness of two regions is fixed. The main idea is to consider one of the arguments (here the second one) as the one determining the scaling context, i.e., the level on the ground of which two regions are defined to be spatially related or not. The results of this chapter can be easily generalised to the case of regions with different types by considering collections of nested partition chains.

With this model in mind, the representation theorem now reads as follows: there is a finite set of axioms such that any binary relation fulfilling these can be represented as a spatial relation based on a nested partition chain. To fit this representation theorem into the pattern of a representation theorem as described in Chap. 2, the instantiations of the template definition are given bellow:

1. The class of structures to be represented *Rnd* is the class of structures of the form $\mathfrak{A} = (pow(X), \delta^{\mathfrak{A}})$ where X is an arbitrary but finite set, $pow(X)$ is the powerset of X and where δ is a binary function on $pow(X)$.
2. The class of representing structures *Rns* is the class of structures of the form $\mathfrak{A} = (pow(X), \mathrm{sr}_{pc})$ where sr_{pc} is a binary relation based on a nested strict partition chain (see below).
3. The underlying construction principle for the representing structures is that of building nested partition chains which present a class of trees with nodes from $pow(X)$.
4. The structure-preserving mapping is that of identity (in particular *Rns* = *Rnd*).

The rest of the chapter is structured as follows. Section 3.2 gives the definitions of partition chains and spatial relatedness. Section 3.3 gives a comparison of spatial relatedness with proximity. Section 3.4 defines the upshift operator used in the axioms. Section 3.5 contains the main axioms for the representation theorem, which, in turn, is proved in Sect. 3.6. The last two sections before the sections on related work and the summary are further extensions of the spatial-relatedness framework: Section 3.7 deals with the question of how to track changes in spatial relatedness when the underlying partition chain changes. In Sect. 3.8, the kind of spatial relatedness considered is not that between arbitrary subsets of a set X, but between regions according to the region connection calculus [4].

The content of this chapter is based on papers [20, 19, 21].

3.2 Partitions and Spatial Relatedness

The binary relations of spatial relatedness for which a representation theorem is going to be developed has as arguments arbitrary subsets of a finite set X. The restriction to finite sets X is relevant for the construction used in the proof of the representation theorem but otherwise the definition and most of the other properties transfer to infinite sets X. A similar assumption of finiteness can be found in the areas of discrete/digital topology [7, 24].

The main technical concept used in the representation theorem is that of a partition and that of a normal partition chain. The usual partition concept of set theory will be called *set partition*. That is, given X and a family of sets $\{a_i\}_{i \in I}$ where the a_is are pairwise disjoint is a set partition iff X is the union of all the a_is, formally $X = \biguplus_{i \in I} a_i$.

Definition 3.1 (partition) A *partition* of a set X on level $i \in \mathbb{N}$ is a family of pairs $(i, a_j)_{j \in J}$ such that $(a_j)_{j \in J}$ is a set partition of X. A pair $c = (i, a_j)$ is called a *cell of level* i. Its underlying set a_j, which is the the second argument of the cell, is denoted $\mathrm{us}(c)$ and its level is denoted $l(c) = i$.

Partition chains are partitions of X that are nested.

Definition 3.2 (partition chain) Consider a collection of $n + 1$ different partitions of X where all partitions have only finitely many cells. This set of partitions is called a *partition chain pc* iff

1. all cells $(i+1, a_j)$ of level $i+1$ (for $i \in \{0, \ldots, n-1\}$) are unions of i-level cells, i.e., there exist (i, b_k), $k \in K$, such that $a_j = \biguplus_{k \in K} b_k$;
2. and the last partition (level n) is made up by (X).

Every cell has a unique upper cell. For a cell (i, a_j) (with $1 \leq i \leq n-1$) let $(i, a_j)^{\uparrow, pc} = (i+1, a_k)$ be the unique cell of the upper level in this partition chain pc such that $a_j \subseteq a_k$. For the cell of level n set $(n, X)^{\uparrow, pc} = (n, X)$. The cell $(i, a_j)^{\uparrow, pc}$ is called the *upper cell* of (i, a_j). If the partition chain is clear from the context, then $(i, a_j)^{\uparrow}$ stands for $(i, a_j)^{\uparrow, pc}$.

A partition chain is *normal* iff all set partitions underlying the partitions are pairwise distinct. A partition chain is *strict* iff for every level i, $i > 0$ and every cell (i, a_j) there is no cell $(i-1, a_j)$ on the level below with the same underlying set.

Example 3.3 An example of a (strict) partition chain with three levels is illustrated in Fig. 3.1, where I give a region oriented presentation (left) and the tree structure (right) with the associated levels. In order to make the example fully concrete I assume that $\text{us}(X) = \{1, 2, 3, 4, 5, 6\}$ and $c_i = (0, \{i\})$, for $i \in \text{us}(X)$.

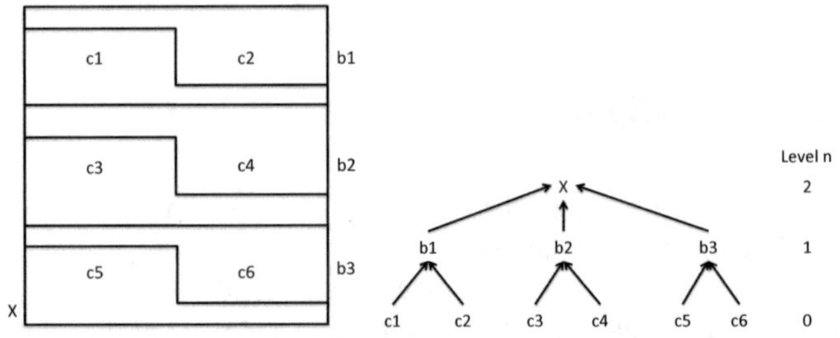

Fig. 3.1 A strict partition chain $(c_i)_{i \in \{1,2,3,4,5,6\}} \leq (b_i)_{i \in \{1,2,3\}} \leq (X)$

For arbitrary subsets $b \neq \emptyset$ of X let $\tilde{b}^{pc}$ denote the cell (i, a_j) such that $b \subseteq a_j$ and i is minimal. The integer $i = l_{pc}(b)$ is called the *level of b in pc*. If the partition chain pc is unique in the used context, it is not mentioned in the subscripts. As a shorthand for $(\tilde{b}^{pc})^{\uparrow, pc}$ one may write $b^{\Uparrow, pc}$.

Example 3.4 Consider again Fig. 3.1. For the set $\{5, 6\} = \text{us}(c_5) \cup \text{us}(c_6)$ it follows that $\widetilde{\{5, 6\}} = b_3$ and so $\{5, 6\}^{\Uparrow, pc} = X$. For the set $\{3, 6\} = \text{us}(c_3) \cup \text{us}(c_6)$ already $\widetilde{\{3, 6\}} = X$ holds, and so again $\{3, 6\}^{\Uparrow, pc} = X$.

Definition 3.5 (spatial relatedness sr) For a normal partition chain pc over X spatial relatedness sr_{pc} is defined by:

$$\text{sr}_{pc}(a, b) \text{ iff } b \neq \emptyset \text{ and } a \cap \text{us}(b^{\Uparrow, pc}) \neq \emptyset$$

So the main idea underlying this definition is that the second argument (here b) determines the partition level w.r.t. which the first argument (here a) is considered to be related. If b is a cell, then one checks whether the intersection of a with the upper cell of b is non-empty. If the intersection is non-empty, then a is spatially related to b, otherwise a is not spatially related to b. If b is not the underlying set of a cell, then one looks for the smallest upper cell whose underlying set contains b and then proceeds as before.

Example 3.6 Consider the partition chain in Fig. 3.2. It is similar to that of Fig. 3.1, but here let $X = \{1, 2, \ldots, 7, 8\}$, $c_i = (0, i)$ for $i \in \{1, 2, 3, 4\}$ and $c_5 = (0, \{5, 7\})$, $c_6 = (0, \{6, 8\})$. Moreover, there is a set $z = \{7, 8\}$ which overlaps with the cells c_5 and c_6 and a set $w = \{6\}$ contained in the cell c_6. It holds that $\tilde{z} = b_3 = (1, \{5, 6, 7, 8\})$ and hence $z^{\Uparrow, pc} = X$. So every set $b \subseteq \mathrm{us}(X)$ is spatially related to z, i.e., $\mathrm{sr}_{pc}(b, z)$. In contrast, consider the set w. Here one has $\tilde{w} = c_6 = (0, \{6, 8\})$ and hence $w^{\Uparrow, pc} = b_3$ $= (1, \{5, 6, 7, 8\})$. So only sets intersecting with $\{5, 6, 7, 8\}$ are spatially related to w.

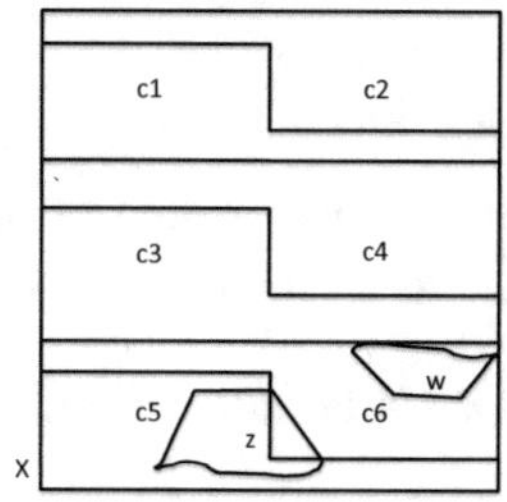

Fig. 3.2 Partition chain with non-cells for illustrating spatial relatedness

Example 3.7 This example illustrates the difference between (metrical) nearness and spatial relatedness. Assume that there are cadastral data covering two different nations and that houses are seen as legal objects adhering to planning laws. Two houses a and b are sited on different sides of the border line of two nations that have two completely different planning laws. Then, the legal object a would not stand in sr_{pc} relation to the legal object b w.r.t. the partition pc made of the cadastral data though the areas a and b are clearly metrically near.

3.3 Spatial Relatedness versus Proximity

A prominent example of qualitative spatial relatedness results from the neighborhood concept of topological spaces. An even more fine-grained mathematical approach to nearness is provided by proximity spaces. These date back to ideas of Riesz presented in a congress talk in 1908 [23]. Proximity spaces were rediscovered in the fifties by the mathematician Efremovič [8, 9]. He gave the axiomatic definition of a proximity space to become the basis for all following work on proximity spaces. I will not delve into the further development of research on proximity spaces but note that proximity spaces also became an important topic in the area of qualitative spatial reasoning [26, 5, 6, 7]. For a historical overview on proximity spaces (until 1970) the reader may have a look at the introductory chapter of the classic monograph by Naimpally and Warrack [18].

In the following, I will not give the definition of proximity spaces according to Efremovič (see [18, p. 7–8]) but rather use the weaker notion of a *minimal proximity relation* given in [7]. The reason is that the spatial relatedness as considered in this monograph is inherently not symmetrical and the total order of partitions is finite, hence induces a discrete approach to nearness which is in the same spirit as the approach of [7].

Structures (X, δ) with domain X and a binary relation δ over X are called *minimal proximity structures* [7] iff the following axioms are fulfilled for all sets $a, b, c \subseteq X$:

(P1) If $\delta(a, b)$, then a and b are non-empty.
(P2*) $\delta(a, b)$ or $\delta(a, c)$ iff $\delta(a, (b \cup c))$.
(P3) $\delta(a, c)$ or $\delta(b, c)$ iff $\delta((a \cup b), c)$.

Proximity spaces are structures that have strong connections to topological spaces. In fact, for a proximity space (X, δ) a canonical topological space $(X, \tau(\delta))$ can be defined by

$$\tau(\delta) = \{A \subseteq X \mid A \text{ is closed according to (3.1)}\}$$

and

$$A \subseteq X \text{ is } \textit{closed under } \delta \text{ iff for all } x \in X\text{: if } \delta(\{x\}, A)\text{, then } x \in A. \tag{3.1}$$

Indeed, $(X, \tau(\delta))$ is a topology in the sense that the following conditions are fulfilled: $\{X, \emptyset\} \subseteq \tau(\delta)$; if $A, B \in \tau(\delta)$, then $A \cup B \in \tau$; and if $(A_i)_{i \in I}$ is a (possibly infinite) family of sets in $\tau(\delta)$, $A_i \in \tau(\delta)$, then $\bigcap_{i \in I} A_i \in \tau(\delta)$. But, as said before, proximity spaces are finer structures than topological spaces in so far as two different proximities δ_1, δ_2 may induce the same topology $\tau(\delta_1) = \tau(\delta_2)$.

It can be easily verified that sr_{pc} fulfils (P1) and (P3), but only the following weakening of (P2*):

(P2) If $\delta(a, b)$ or $\delta(a, c)$, then $\delta(a, (b \cup c))$

A counterexample for (P2*) is discussed later in Example 3.22. Moreover one can show that sr_{pc} fulfils the following two properties:

(P4) If $a \cap b \neq \emptyset$, then $\delta(a, b)$ and $\delta(b, a)$.
(P5) For all $a \subsetneq X$ with $a \neq \emptyset$: $\delta(a, (X \setminus a))$ or $\delta((X \setminus a), a)$.

The following proposition summarises these results.

Proposition 3.8 *All sr_{pc} for normal partition chains pc fulfill the axioms (P1), (P2), (P3), (P4), and (P5).*

Proof See p. 66. □

As the following example shows, this set of axioms is incomplete in the following sense: there are still models where δ is interpreted by a binary relation that cannot be represented as sr_{pc} for an appropriate partition chain pc. In other words, these axioms do not completely characterize the relations of the type sr_{pc}.

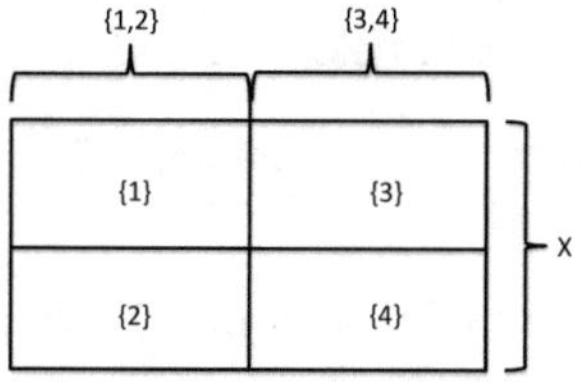

Fig. 3.3 A counter example

Example 3.9 Assume $X = \{1, 2, 3, 4\}$ and the δ-relations fulfill the following conditions (see also Fig. 3.3):

- None of the following holds: $\delta(\{4\}, \{2\})$, $\delta(\{4\}, \{1\})$, $\delta(\{1\}, \{3\})$, $\delta(\{1\}, \{4\})$
- for all other $a, b \subseteq X$ with $a, b \neq \emptyset$ it holds that $\delta(a, b)$.

It can be easily checked that δ fulfills (P1)–(P5), but that it is not representable as sr_{pc} for a normal partition chain.

The last assertion is proved as follows: Take the assertion $\delta(3, 2)$. Assume that there is a normal pc such that $\delta = \mathrm{sr}_{pc}$. Consider the following cases:

1. $c := \widetilde{\{2\}} = (0, \{2\})$. As $\delta(\{1\}, \{2\})$ and $\delta(\{3\}, \{2\})$ it must be the case that $\{1, 2, 3\} \subseteq \mathrm{us}(c^{\uparrow})$. As not $\delta(\{4\}, \{2\})$, $\mathrm{us}(c^{\uparrow}) = \{1, 2, 3\}$. That means that on level 1 one can have only the sets $\{1, 2, 3\}$ and $\{4\}$ as underlying cells. But this means that $\widetilde{\{4\}} = (0, \{4\})$ and $4^{\Uparrow} = (1, \{4\})$. But this contradicts the fact that $\delta(\{2\}, \{4\})$ holds while one would have to have not $\mathrm{sr}_{pc}(2, 4)$.
2. In the other cases $c := \widetilde{\{2\}} = (0, a)$ for a set a with $\{2\} \subsetneq a$. But then $c^{\uparrow} = (1, b)$ for a set b which must again be $b = \{1, 2, 3\}$ for the same reasons as in the former case. But then one gets a contradiction again.

3.4 The Upshift Operator

The main idea for the representation theorem is to reconstruct the levels by referring only to δ. A first step towards this end is to defimyExamplene the *upshift operator* $\cdot^{\Uparrow_\delta}$ *for* δ, an abstract analogue of the level-shifting operator $\cdot^{\Uparrow, pc}$. I refer to this abstract operator also under the term *δ-upshift*. The upshift operator is going to be defined below as a unique function based on δ. In all axioms where $\cdot^{\Uparrow_\delta}$ occurs it can be unfolded to its defining formula to get rid of the new symbol.

Given δ, the equivalence relation ${}^{\bullet}\!\sim$ is defined as follows:

$$a \;{}^{\bullet}\!\sim\; b \text{ iff } \{c \subseteq X \mid \delta(c, a)\} = \{c \subseteq X \mid \delta(c, b)\} \tag{3.2}$$

This equivalence relation can be formulated for any relation δ, independently of the specific properties of δ. As usual, for any equivalence relation $\sim$, $[a]_\sim$ denotes the equivalence class of a w.r.t. $\sim$. A simple observation is the following:

Proposition 3.10 *For partition chains pc and $a, b \subseteq X$ s.t. $\tilde{a} = (i, a)$, $\tilde{b} = (i, b)$, and $a^{\Uparrow, pc} = b^{\Uparrow, pc}$ it holds that $a \;{}^{\bullet}\!\sim\; b$.*

Definition 3.11 Given a binary relation δ, the *upshift operator* $\cdot^{\Uparrow_\delta}$ *for* δ is defined for any non-empty set $b \subseteq X$ as follows:

$$b^{\Uparrow_\delta} = \bigcup [b]_{{}^{\bullet}\!\sim}$$

So, the set $b^{\Uparrow_\delta}$ is just the union of all sets a that have the same set of sets that are δ-related to them as b. If a partition chain pc over X is given, then one has two different shift operators, the level-shift operator $\cdot^{\Uparrow, pc}$, which calculates the upper cell w.r.t. pc, and the upshift operator $\cdot^{\Uparrow_{\mathrm{sr}_{pc}}}$ w.r.t. sr_{pc}. But, as the following proposition shows, the δ-upshift operator is nothing else than the level-shift operator in case of $\delta = \mathrm{sr}_{pc}$.

Proposition 3.12 *Let pc be a partition chain over X. Then for any non-empty $b \subseteq X$ the following equality holds:* $b^{\Uparrow, pc} = b^{\Uparrow_{\mathrm{sr}_{pc}}}$.

3.5 Main Axioms

In the following subsections the main axioms that lead to the representation theorem are introduced and discussed.

3.5.1 Spatial Relatedness is Grounded

The following axiom states a necessary and sufficient condition for the spatial relatedness of two sets with reference to the upshift operator $\cdot^{\Uparrow_\delta}$. It says that a is δ-related to b if and only if a has a non-empty intersection with the upshift of b.

(Pgrel) For all $a, b \subseteq X$: $\delta(a, b)$ iff $a \cap b^{\Uparrow_\delta} \neq \emptyset$.

The axiom expresses a principle on the connection between the abstract δ relation and the set-theoretic element-of relation $\in$: namely that δ is **gr**ounded in the **el**ement relation (hence the acronym grel).

Unfolding (Pgrel) w.r.t. the definition of $\cdot^{\Uparrow_\delta}$ results in the axiom (Pgel'):

(Pgrel') For all $a, b \subseteq X$: $\delta(a, b)$ iff there is some c such that $c \overset{\bullet}{\sim} b$ and $a \cap c \neq \emptyset$.

Further, the relation symbol $\overset{\bullet}{\sim}$ can be eliminated—leading to (Pgrel") which refers only to δ (and some set operations).

(Pgrel") For all $a, b \subseteq X$: $\delta(a, b)$ iff there is some c such that for all z: $\delta(z, c)$ iff $\delta(z, b)$, and $a \cap c \neq \emptyset$.

Intuitively speaking, (Pgrel") says that a is δ-related to b iff it has a non-empty intersection with a set c that is similar ($\overset{\bullet}{\sim}$ equivalent) to b. Looking at the unfolding, it is no surprise that (Pgrel) on its own is not expressive enough to characterise sr_{pc} and hence is far away from being a definition of sr_{pc}. This is demonstrated by Example 3.13.

Example 3.13 Let $X = \{1, 2, 3\}$ and δ be as follows:

- for all $a \subseteq X$: $\delta(a, \{1\})$
- $\neg\delta(1, 2)$
- $\neg\delta(1, 3)$

One calculates $2^{\Uparrow_\delta} = 3^{\Uparrow_\delta} = \{2, 3\}$ and $\{2, 3\}^{\Uparrow_\delta} = X$ and shows that (Pgrel) is fulfilled. Nonetheless this δ is not representable as sr_{pc} for some normal pc.

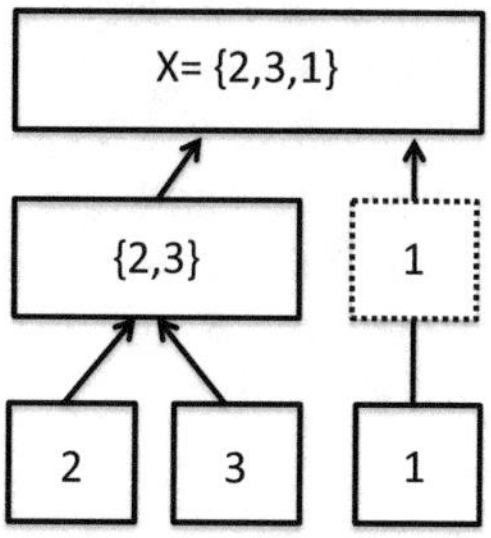

The reason for non-representability in the above example is that there is no appropriate level notion. All of the sets $\{1\}$, $\{2\}$, and $\{3\}$ would have to be of level 0. Applying $\cdot^{\Uparrow_\delta}$ to $\{2\}$ and $\{3\}$ gives $\{2, 3\}$, but the application to $\{1\}$ already gives X. Hence $\{1\}$ would have to appear on two levels (serving also as a cell on the level of $\{2, 3\}$), but this would mean that the only set δ-related to $\{1\}$ is X—which is not the case.

Nonetheless, (Pgrel) has some consequences for the other axioms. In order to give a more detailed view, (Pgrel) is divided into two sub-axioms.

(Pgreln) For all $a, b \subseteq X$: if $\delta(a, b)$, then $a \cap b^{\Uparrow_\delta} \neq \emptyset$.

The added "n" stands for "necessary condition" because a necessary condition is specified for δ.

(Pgrels) For all $a, b \subseteq X$: if $a \cap b^{\Uparrow_\delta} \neq \emptyset$, then $\delta(a, b)$.

The "s" stands for sufficient condition.

Proposition 3.14 *The following entailment relations hold:*

1. *(Pgreln), (Pgrels)* $\models$ *(P3)*
2. *(Pgrels)* $\models$ *(P4)*

Proof See p. 67. □

So, with (Pgreln) and (Pgrels) the axiom (P3) becomes redundant, and (Pgrels) already entails (P4). (Pgrels) is already entailed by (P4). And hence (Pgrels) and (P4) are equivalent.

Proposition 3.15 *(P4)* $\models$ *(Pgrels)*

Proof See p. 67. □

A simple consequence of axioms (Pgreln), (P2), (P4) is the monotonicity of the upshift operator.

Proposition 3.16 *If the axioms (Pgreln), (P2), (P4) hold, then monotonicity holds: for all* $a \subseteq b$ *one has* $a^{\Uparrow_\delta} \subseteq b^{\Uparrow_\delta}$.

Proof See p. 67. □

So, the upshift operator $\cdot^{\Uparrow_\delta}$ fulfills one of the conditions of a closure operator in a topological sense. But $\cdot^{\Uparrow_\delta}$ is not a closure operator, not even a pre-closure/Cech-operator, i.e., it does not fulfil the following conditions for an operator $f : pow(X) \longrightarrow pow(X)$: (i) $f(\emptyset) = \emptyset$; (ii) $a \subseteq f(a)$; (iii) $f(a \cup b) = f(a) \cup f(b)$. Here, as in the whole monograph, $pow(X)$ stands for the power set of X. Condition (i) is not fulfilled as it is not defined for empty sets—but this could be remedied. Condition (ii) is fulfilled (under (P4)), but Condition (iii) states distributivity w.r.t. the union of sets.

3.5.2 Alignment of Upshift Close-ups

The upshift operator is intended to produce cells only. One aspect of this property is captured by the following nestedness condition.

(Pnested) For $a, b \subseteq X$: either $a^{\Uparrow_\delta} \subseteq b^{\Uparrow_\delta}$ or $b^{\Uparrow_\delta} \subseteq a^{\Uparrow_\delta}$ or $a^{\Uparrow_\delta} \cap b^{\Uparrow_\delta} = \emptyset$.

As mentioned above, $\cdot^{\Uparrow_\delta}$ is not a closure operator. Nonetheless, one can state the following axioms characterising the behaviour of the double application of the operator—replacing idempotence—and characterising the outcome of applying it to a union of sets—replacing distributivity over unions of sets.

Axiom (Pdoubleshift) states that if the upshift of a is properly contained in a cell (the upshift of b), then another upshift application will keep it in this cell.

(Pdoubleshift) If $a^{\Uparrow_\delta} \subsetneq b^{\Uparrow_\delta}$, then $a^{\Uparrow_\delta \Uparrow_\delta} \subseteq b^{\Uparrow_\delta}$.

Axiom (Punionshift) determines the upshift of the union of two sets a and b under the condition that double applications lead to the same set and that the outcomes of single applications are not comparable: in this case, the upshift of the union is the same as applying the upshift three times to a or b.

(Punionshift) If $a^{\Uparrow_\delta \Uparrow_\delta} = b^{\Uparrow_\delta \Uparrow_\delta}$ and $a^{\Uparrow_\delta} \not\subseteq b^{\Uparrow_\delta}$ and $b^{\Uparrow_\delta} \not\subseteq a^{\Uparrow_\delta}$, then

$$(a \cup b)^{\Uparrow_\delta} = a^{\Uparrow_\delta \Uparrow_\delta \Uparrow_\delta} = b^{\Uparrow_\delta \Uparrow_\delta \Uparrow_\delta}.$$

Proposition 3.17 *All sr_{pc} over a normal partition chain pc fulfill (Pdoubleshift) and (Punionshift).*

Proof See p. 67. □

The axioms above do not capture the effect of the $\tilde{\cdot}$ operator, which makes spatial relatedness being determined by its underlying cells. The main observation here is given by the following axiom. Intuitively, it says that subsets of two sets which are not upshift comparable lead to the same upshift.

(Pcelldet) If $a^{\Uparrow_\delta} \not\subseteq b^{\Uparrow_\delta}$ and $b^{\Uparrow_\delta} \not\subseteq a^{\Uparrow_\delta}$, then for all $a' \subseteq a$ and $b' \subseteq b$ (with $a', b' \neq \emptyset$) it follows that $(a' \cup b')^{\Uparrow_\delta} = (a \cup b)^{\Uparrow_\delta}$.

Proposition 3.18 *All sr_{pc} over a normal partition chain pc fulfill (Pcelldet).*

Proof See p. 68. □

3.5.3 Isolated Points

An interesting point regarding $\cdot^{\Uparrow_\delta}$ is that it may contain fixed points or *isolated points*—as they are denoted in the following. In fact, for normal partition chains in which you may have sets a that occur on more than one level, lets call them *pc-fixed points*, it holds that $a^{\Uparrow_{\mathrm{sr}_{pc}}} = a$: $a = a^{\Uparrow,pc} \overset{(Prop.\ 3.12)}{=} a^{\Uparrow_{\mathrm{sr}_{pc}}}$.

Definition 3.19 (upshift-isolated) A set $a \subseteq X$ is upshift isolated, for short $\mathsf{uiso}(a)$, iff $a^{\Uparrow_\delta} = a$.

Now let us look again at points a in a normal partition pc that are pc-fixed points. Another property these sets have is the following: if $\mathrm{sr}_{pc}(x, a)$, then $a \cap x \neq \emptyset$. Hence one may define the following equivalent notion of isolation:

Definition 3.20 (set-isolated) A set $a \subseteq X$ is set-isolated, for short: $\mathsf{siso}(a)$, iff: for all $x \subseteq X$: if $\delta(x, a)$, then $x \cap a \neq \emptyset$.

A simple observation is that these notions are the same if (Pgreln) and (P4) are fulfilled.

Proposition 3.21 $(Pgreln), (P4) \models \forall a.\mathsf{uiso}(a) \leftrightarrow \mathsf{siso}(a)$.

Proof See p. 68. □

3.5.4 Splittings

In general, sr_{pc} relations do not fulfil the other direction in axiom (P2*) which states that if a is δ-related to $b \cup c$, then a is δ-related to b or c. The reason is that $b \cup c$ gives a coarser scaling context spatial relatedness than does any of its parts b or c. This is in the very nature of scaled proximity and is illustrated in the following example.

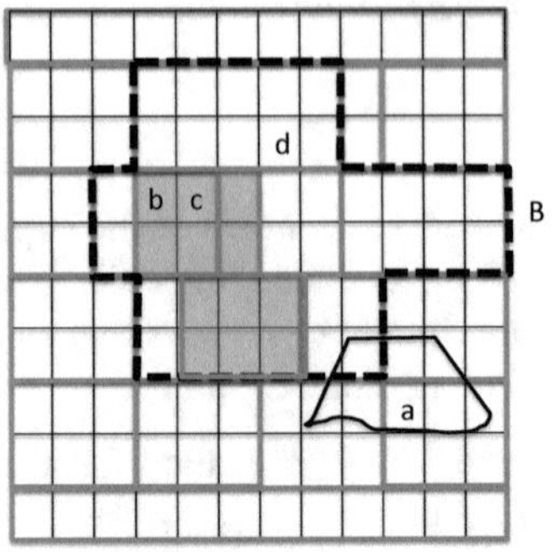

Fig. 3.4 Counterexample to right additivity

Example 3.22 Consider the partitions illustrated in Fig. 3.4. In Fig. 3.4 the smallest rectangles represent the finest partition, two of them being b, c. The next upper level consists of cells represented with grey border lined rectangles, in particular d is one. The region B (dashed border line) is a cell of the third level. As one can see, a is spatially related to $b \cup c$, i.e., $\mathrm{sr}_{pc}(a, b \cup c)$, because a has a non-empty intersection with $B = (b \cup c)^{\Uparrow, pc}$. But a is not spatially related to any of the six cells (in particular not to b or c) that make up d, because a does not intersect with $d = (b)^{\Uparrow, pc} = (c)^{\Uparrow, pc}$.

Following [21], I call the pair (b, c) with $b \cap c = \emptyset$ an *irregular split* of $b \cup c$ w.r.t. a. The main observation is that for any a there can be at most one irregular split.

Proposition 3.23 *For sr_{pc}, every set a has at most one irregular split.*

Proof See p. 68. □

This property will now be formulated as an axiom over δ:

(PirrSplit) For δ, every a has at most one irregular split.

Any relation δ fulfilling (P2) and (PirrSplit) has a partition of X into cells which can serve as the cells of level 0. The crucial concept is the following.

Definition 3.24 (cell-equivalence) For all $x, y \in X$ let the relation of cell-equivalence, $\sim_0$ for short, be defined by

$$x \sim_0 y \text{ iff } \{x\} \;{}^{\bullet}\!\!\sim \{x, y\} \text{ and } \{y\} \;{}^{\bullet}\!\!\sim \{x, y\}$$

The cell-equivalence relation is indeed an equivalence relation:

Proposition 3.25 *Assume δ fulfills (P2) and (PirrSplit). Then the relation $\sim_0$ is an equivalence relation, i.e., it is symmetric, transitive, and reflexive.*

Proof See p. 69. □

Actually, using the same proof idea, it is possible to prove the following theorem, which generalizes the result of the proposition.

Theorem 3.26 *For all subsets* $b_1, b_2 \subseteq [x]_{\sim_0}$: $b_1 \;^{\bullet}\!\!\sim b_2$.

Proof See p. 69. □

So, this result gives the base on which to build the partition chain, namely the zero-level partition consisting of cells $[x]_{\sim_0}$.

3.6 Representation Theorem for Spatial Relatedness

This section gives the proof for the representation theorem for those spatial relations that are based on strict partition chains. A problem on building further cells upon cells $[x]_{\sim_0}$ are isolated sets. Hence, isolated sets are explicitly excluded.

(Pnoiso) For every $a \subsetneq X$ one has: $a \neq a^{\Uparrow_\delta}$.

The main problem in representing spatial relatedness is to capture the fact that all paths from the root to the leaves in the pc have the same length. So I consider the following notion of rank for any binary relation δ on X.

Definition 3.27 (rank) For any $a \in pow(X) \setminus \{\emptyset\}$ define by induction on $n \in \mathbb{N}$: $a^0 = a$ and $a^{n+1} = (a^n)^{\Uparrow_\delta}$. Then the *rank* of a is:

$$r(a) = \begin{cases} m & \text{s.t. there is } m' \text{ with } a^{m'} = X \text{ and} \\ & m \text{ is the minimal one amongst the } m' \\ \infty & \text{else} \end{cases}$$

The second case comes into play when there are isolated sets.[1] Now one can formulate the following axiom which says that every pair of singleton sets $\{x\}$, $\{y\}$ over the domain X have the same rank

(Psamerank) For all $x, y \in X$: $r(\{x\}) = r(\{y\})$.

Due to the definition of a strict partition chain the following holds.

Proposition 3.28 sr_{pc} *over a strict partition chain pc fulfills (Psamerank).*

With these additional axioms, the representation theorem for spatial relations generated by strict partition chains can be proved.

Theorem 3.29 *If* δ *fulfills (P1), (P2), (Pgreln), (Pgrels), (Pnoiso), (PirrSplit), (Psamerank), (Pcelldet), (Pdoubleshift), and (Punionshift), then there is a strict pc, such that* $\delta = sr_{pc}$.

[1] Though axiom (Pnoiso) excludes isolated sets, the definition of rank accounts for them. This is motivated by the aim of keeping all axioms, in particular (Pnoiso) and (Psamerank), independent of each other.

Proof See p. 70. □

Together with the propositions proved before one gets the following corollary.

Corollary 3.30 *A binary relation δ fulfills (P1), (P2), (Pgreln), (Pgrels), (Pnoiso), (PirrSplit), (Psamerank), (Pcelldet), (Pdoubleshift), and (Punionshift) if and only if there is a strict partition chain pc, such that* $\delta = sr_{pc}$.

3.7 Dynamics of Partition Chains

Until now I assumed that the partition chain pc underlying the spatial relatedness relation sr_{pc} is fixed, i.e., does not change (in time). This does not reflect in full the reality concerning cadastral data. Hence, in this short section, I consider change operators on partition chains and investigate the question how this change effects the induced spatial relatedness. More concretely, assume that a partition chain pc_1 is changed to a new partition chain pc_2. What can be said about the change from the induced spatial relatedness sr_{pc_1} to the induced spatial relatedness sr_{pc_2}? In particular, one can ask what kind of change transitions $pc_1 \leadsto pc_2$ do not change the spatial relatedness, $\mathrm{sr}_{pc_1} = \mathrm{sr}_{pc_2}$, or between what sets on what level does a change of the total orderings affect the nearness in between them. Similar problems have been tackled by [13] and especially [25], which considers the global dynamics of tree-like spatial configurations.

The change transition $\leadsto$ between total orders are not allowed to be arbitrary transitions but some intuitive changes which have corresponding real world counterparts. In particular, the kind of changes that are worth being investigated are the merger of regions, the switch of levels, the additions of partitions etc.

Investigations into this kind of relation are necessary for a formal theory of dynamics of nearness. In particular such a theory provides a formal grounding for optimizations within a cognitive agent that bases its notion of spatial relatedness on partition chains; rather than re-calculating the spatial relatedness between all regions in case the agent moves around (local change) or a partition chain is updated (global change) it directly uses the knowledge on regions between which spatial relatedness is expected to have changed. This section describes the first ideas on a foundation for such a theory.

In their study of regional changes of municipalities in Finland, Kauppinen and colleagues [14] found seven kinds of type changes which are as follows: 1. A region is established. 2. Two or more regions are merged into one. 3. A region is split into two or more regions. 4. The name of a region is changed. 5. A region is annexed to a different country. 6. A region is annexed from a different country. 7. A region is moved to another city or municipality. I am interested in changes that concern changes of cells for partitions in a given partition chain. Hence I adapt a subset of the types of changes to my setting by explicitly formalizing the type of change.

Clearly the most interesting changes are that of merging two regions to a new region and its counterpart, the split of regions into two regions. These types of

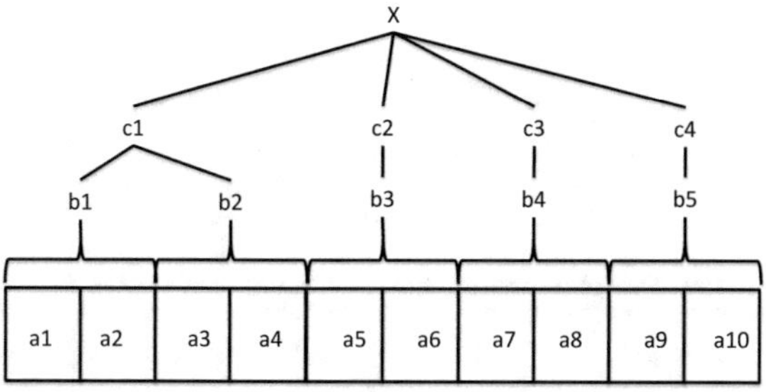

Fig. 3.5 Illustration of example configuration for merge

changes are low frequent-changes (in contrast to the local dynamics case where an agent updates the nearness relations when moving around). For example, Kauppinen and colleagues [14] recognized 144 merges and 94 splits of municipalities in Finland between 1865 and 2007. But nonetheless, the effects of merges and splits on spatial relatedness are worth to be investigated.

Here, I restrict my attention to different forms of merging. I have to explain what it means that two cells (of a partition) are merged, and whether such a merge is possible such that the result is again a (normal) partition chain.

Let pc be a normal partition chain over X having levels 0 to n. I will look at merging two cells on the same level into a new cell. In order to get a first rough picture on the effects of merging, I look at the special case where the cells are members of the next-to-last level $n - 1$. In this case, both cells to be merged have always the same upper cell, namely X. For illustration of the possible merge operations have a look at the partition chain in Fig. 3.5, which I have arranged such that one can see the tree structure of the the partition chain, with X being its root. The cells labelled with the letter c make up the cells of the next-to-last level 2. The different forms of changes within a partition chain can be seen as different forms of updating a tree.

Merging the cells $(2, c_2)$ and $(2, c_3)$ into a new cell means that the underlying set of the merging result has to have the union of c_2 and c_3 as the underlying set. But there are in principle two ways to conduct this merge that depend on specifying the level of the merge result.

The first option is to modify the next-to-last level, so that the whole number of levels is untouched. In case of the example illustrated in Fig. 3.5 this would mean that the partition of c-cells is substituted by the new partition of c-cells that consists of the cells $(2, c_1)$, $(2, c_2 \cup c_3)$ and $(2, c_4)$ (see Fig. 3.6). I term this type of merge *level modifying merge—lm merge* for short. If a normal partition chain pc_2 results from another normal partition chain pc_1 by an lm merge, then I write $pc_1 \leadsto^{lm} pc_2$.

The other option is to make the union of the sets to be part of a new level. Hence, in addition to the original partition made up by $(2, c_1)$, $(2, c_2)$, $(2, c_3)$ and $(2, c_4)$, one adds the partition $(3, c_1)$, $(3, c_2 \cup c_3)$ and $(3, c_4)$ and raises the level of X by one to $(4, X)$ (see Fig. 3.7). I term this type of change *level adding merge—la merge* for

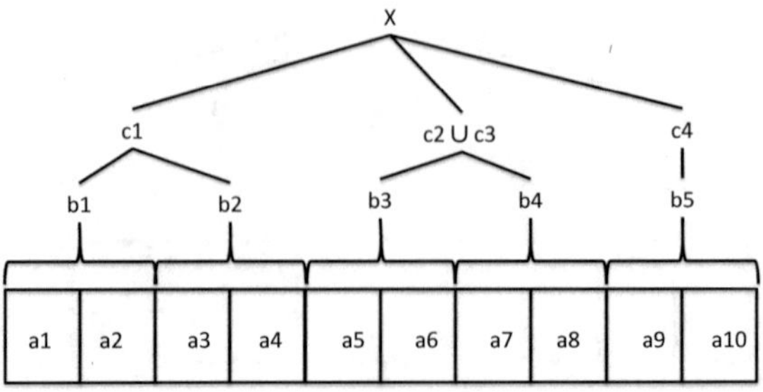

Fig. 3.6 Illustration of merge by operation of type modifying

short. If a normal partition chain pc_2 results from another normal partition chain pc_1 by an la merge, then I write $pc_1 \rightsquigarrow^{la} pc_2$.

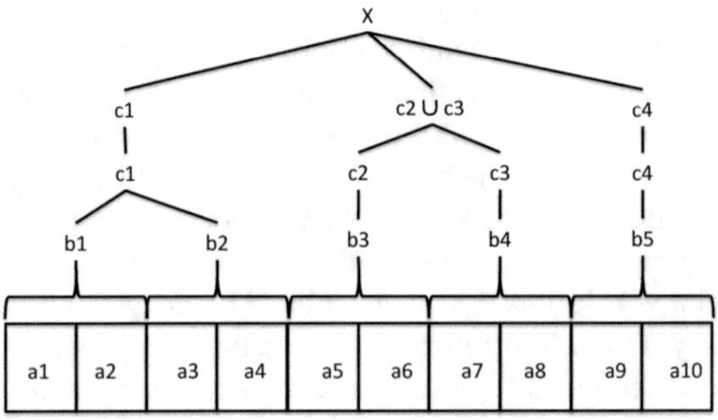

Fig. 3.7 Illustration of merge by operation of type adding

In some cases, neither form of merge may be possible without violating the normality condition. For example, if the next-to-last level consists only of two cells $(n-1, x_1)$ and $(n-1, x_2)$, then the union of x_1 and x_2 is the whole domain X.

What can one say about the change of spatial relatedness induced by level modifying merges on the next-to-last level? First, note that the level of a set in pc_1 is identical to the level in pc_2 if the former is below or equal to $n-1$. If its level in pc_1 is n, then its level in pc_2 may be n or $n-1$.

The change of pc_1 into pc_2 affects only the next-to-last partition, e.g., by merging cells $(n-1, c_1)$ and $(n-1, c_2)$. Hence, spatial relatedness is affected only locally. So, if the second argument b has level at most $n-3$, then one can say that a is spatially related to b in pc_2 if and only if it is spatially related in pc_1. So one can prove the following proposition.

Proposition 3.31 *Let pc_1, pc_2 be two normal partition chains over X such that $pc_1 \rightsquigarrow^{lm} pc_2$ w.r.t. cells $(n-1, c_1)$ and $(n-1, c_2)$ on the next-to-last level $n-1$. Then the following assertions hold:*

1. *For all sets $a \subseteq X$ and all sets $b \subseteq X$ with level $l_{pc_2}(b) \leq n-3$ one has: $sr_{pc_1}(a, b)$ iff $sr_{pc_2}(a, b)$.*
2. *For all sets $a, b \subseteq X$: if $sr_{pc_1}(a, b)$, then $sr_{pc_2}(a, b)$.*

Proof See p. 71. □

The consequence of this proposition for a cognitive agent using sr as his concept of spatial relatedness is that it has to update his sr graph only locally when the partition chain is updated by a level modifying change.

Due to the level addition, the situation for la merges is a little bit different. For example, considering our example partition chain illustrated in Fig. 3.5 one can have $a \subseteq X$ such that $\text{sr}_{pc_1}(a, c_2)$ but not $\text{sr}_{pc_2}(a, c_2)$, because the upper level cell of $(n-1, c_2)$ in pc_1 is the biggest cell (n, X), but in pc_2 the upper cell is $(n-1, c_2 \cup c_3)$. So, choosing, e.g., $a = a_1$ and $b = c_2$ one has $\text{sr}_{pc_1}(a, b)$ but not $\text{sr}_{pc_2}(a, b)$. But still it can be shown as above that sets with level below $n-3$ have the same spatial relations.

Proposition 3.32 *Let pc_1, pc_2 be two normal partition chains over X such that $pc_1 \rightsquigarrow^{la} pc_2$ w.r.t. cells $(n-1, c_1)$ and $(n-1, c_2)$ on the next-to-last level $n-1$. Then for all sets $a \subseteq X$ and all sets $b \subseteq X$ with level $l_{pc_2}(b) \leq n-3$ one has : $sr_{pc_1}(a, b)$ iff $sr_{pc_2}(a, b)$.*

3.8 Spatial Relatedness for Region-Based Calculi

The general idea of spatial relatedness can be extended from pure (finite) sets to regions defined according to axioms in a qualitative spatial calculus. Here I demonstrate the extension for the region connection calculus [22] described also in Chap. 2 of this monograph.

A small example will show the kind of generalisations that are required.

Example 3.33 As mentioned before, natural examples of nested partitions are cadastral data. Figure 3.8 shows part of a map of Switzerland with cadastral data with three levels of administrative regions: the most fine-grained level of municipalities, the upper level of districts and the roughest level of cantons. Of course, one has to generalise the notion of partition in order to incorporate the borders of regions. But the generalisation is straight-forward when using the summation notion based on the RCC8 calculus (see [19]): the regions in a partition level have to cover the whole space and are allowed to touch each other.

Now, if one starts a geographical query search on regions that are spatially related to the municipality Dietlikon, i.e., one asks for a such that $\text{sr}_{pc}(a, dietlikon)$ then according to the definition of sr_{pc} one has to consider the upper administrative region

Fig. 3.8 Administrative regions spatially related to Dietlikon

of Dietlikon, which is Bülach: then every region having a non-empty intersection with Bülach (also those touching it) are considered to be spatially related to Dietlikon. In particular all named administrative regions in Fig. 3.8 are spatially related to Dietlikon. For example, Opfikon is spatially related because it is strictly contained in Bülach, and also Rümlang because it touches Bülach.

The generalisation of the notion of a partition chain is straight forward: given a region X, an RCC partition $(a_i)_{i\in I}$ over X is a family of RCC-regions a_i such that:

1. $\bigcup_{i\in I} a_i = X$, where $\bigcup_i$ is the more convenient notation for the RCC summation of regions, and
2. if $i \neq j$ then $a_i\{\mathsf{DC}, \mathsf{EC}\}a_j$, that is, the a_i are pairwise discrete.

The a_i are called *cells*. Note that here I do not explicitly state the level of the cell. The partition is called finite if the index set I is finite.

The notion of an RCC partition chain is then again that of a total ordering of nested RCC partitions where the last (roughest) RCC partition is the region X. In the sections before I already used this notion of RCC partition chains when visualising them as regions in the two dimensional area. In fact, the partition chain example of Fig. 3.1 is equally (or more properly) an example for an RCC partition chain.

The RCC partition chain of Fig. 3.1 is symbolically represented as a set of assertional axioms as follows (where $X, c_1, \ldots, c_6, b_1, b_2, b_3$ denote constants and *cell* is a unary predicate symbol denoting regions appearing as cells in a partition chain.).

$$X = b_1 \cup b_2 \cup b_3 \wedge b_1\{\mathsf{ec}\}b_2 \wedge b_1\{\mathsf{dc}\}b_3 \wedge b_2\{\mathsf{ec}\}b_3$$
$$b_1 = c_1 \cup c_2 \wedge b_2 = c_3 \cup c_4 \wedge b_3 = c_5 \cup c_6$$
$$c_1\{\mathsf{ec}\}c_2 \wedge c_1\{\mathsf{ec}\}c_4 \wedge c_3\{\mathsf{ec}\}c_4 \wedge c_3\{\mathsf{ec}\}c_6 \wedge c_6\{\mathsf{ec}\}c_5$$
$$c_2\{\mathsf{dc}\}c_3 \wedge c_2\{\mathsf{dc}\}c_4 \wedge c_2\{\mathsf{dc}\}c_5$$
$$c_2\{\mathsf{dc}\}c_6 \wedge c_1\{\mathsf{dc}\}c_3 \wedge c_1\{\mathsf{dc}\}c_5$$
$$c_1\{\mathsf{dc}\}c_6 \wedge c_4\{\mathsf{dc}\}c_5 \wedge c_4\{\mathsf{dc}\}c_6 \wedge c_3\{\mathsf{dc}\}c_5$$
$$\forall x[cell(x) \leftrightarrow (x = X \vee x = b_1 \vee x = b_2 \vee x = b_3$$
$$\vee x = c_1 \vee \cdots \vee x = c_6)]$$

For every partition chain pc, $\mathcal{A}_{pc}$ or even shorter $\mathcal{A}$ denotes the set of axioms representing the partition chain with the sorted predicate logic used for BRCC.

Again, as before one may restrict the notions of partition chains to those that are normal and even more to those that are strict.

In real-world applications partitions can be safely assumed to be normal as otherwise a distinction between the administrative units would not be introduced. But, as in the case of normal partition chains, it may be the case that the same region has two different administrative functions.

Example 3.34 I give an example of a normal but non-strict RCC partition chain in Fig. 3.9. Here the regions b_1 and c_1 are objects with the same local extension, but with different levels (c_1 being the lowest level and b_1 being on the next upper level). Nonetheless, the partition is normal, because all partitions on the three levels are different.

Note that a given partition (a) with i different cells induces a *set* of normal partition chains, more specifically, a lattice where the partitions are partially ordered by the degree of granularity.

For the definition of RCC spatial relatedness I follow the three-step approach given in [19] which in turn extended the two-step approach of [12]. In all of the following definitions let $(a) = (a_i)_{i \in I}$ denote a finite partition of the normal partition chain. In the first step, spatial relatedness is defined between regions x_a, y_a on the same level (i.e. from the same partition (a) as denoted by the subscripts), resulting in *a priori spatial relatedness*. In the second step, *a priori* spatial relatedness is used to

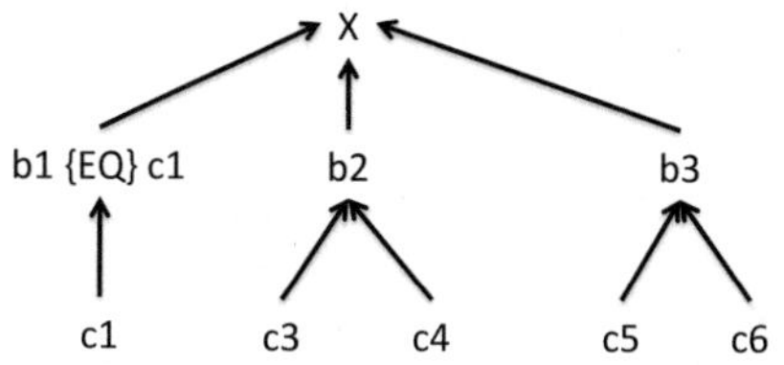

Fig. 3.9 Normal partition chain with cells b_1, c_1 having same local extension

define spatial relatedness between any region z as first argument and a cell as second argument. Last, in the third step the definition is extended to any pair of regions.

One reason for a three-step construction is, firstly, that of a better comprehension of the definitions. But there are two other reasons: The first is to enable talking about different of spatial relatedness not depending on the local extension of objects. For example, it may be the case that one is interested in spatial relatedness to Hamburg considered as a city on the one hand or to Hamburg considered as a Bundesland (federal state) on the other hand. According to the approach in the section before it would be possible to talk only about spatial relatedness w.r.t. the city Hamburg which is on the finer level than the federal state of Hamburg. Rather than defining spatial relatedness for the spatial extension (the underlying set) of a cell one would have to incorporate its level. This is implicitly done for the definition of a-priori spatially relatedness where the level of partition is explicitly given.

The next reason for using a multi-step approach, which is actually related to the first one, comes from the idea of training an agent to extend his knowledge on spatial relatedness between regions: first he learns only spatial relatedness between regions on the same level and then extends this knowledge to regions of different levels. A detailed explanation of this idea is out of the scope of this monograph as it would have to discuss in detail an epistemic logic to formalise the knowledge of agents.

A priori relatedness between two regions x_a, y_a of the same level is defined to hold iff they are part of the same upper cell b:

$$\forall x_a \in (a)\ \forall y_a \in (a)\ [\text{rcc-sr}_{\text{ap}}(y_a, x_a) \text{ iff } \exists b \in (a)^{\uparrow}[\mathsf{P}(x_a, b) \wedge \mathsf{P}(y_a, b)]] \quad (3.3)$$

Example 3.35 Consider the partition chain of Fig. 3.1. The only cells x_a that are *a priori* spatially related to region c_2, formally: $\text{rcc-sr}_{\text{ap}}(x_a, c_2)$, are $x_a = c_1$ and $x_a = c_2$, as these are the only regions contained in the same cell b_1 as c_2.

Spatial relatedness between regions z, which are not necessarily cells, and cells x_a is defined by the following condition saying that there is a region y_a (on the same level as x_a) *a priori* spatially related to x_a such that z and y_a are connected. Again, the second argument determines the scaling context (the level) w.r.t. which spatial relatedness is calculated.

$$\forall x_a \in (a)\ \forall z\ [\text{rcc-sr}(z, x_a) \text{ iff } \exists y_a \in (a)[\text{rcc-sr}_{\text{ap}}(y_a, x_a)) \wedge \mathsf{C}(z, y_a)]] \quad (3.4)$$

Expanding (3.4) with the definition for $\text{rcc-sr}_{\text{ap}}$ in (3.3) leads to the following condition:

$$\forall x_a \in (a) \forall z[\text{rcc-sr}(z, x_a) \text{ iff } \exists b \in (a)^{\uparrow} \exists y_a \in (a)[\mathsf{P}(x_a, b) \wedge \mathsf{P}(y_a, b) \wedge \mathsf{C}(z, y_a)]]$$

The expression on the right hand side can be reduced according to the definition of P in Sect. 2.4 from $\exists b \in (a)^{\uparrow} \exists y_a \in (a)[\mathsf{P}(x_a, b) \wedge \mathsf{P}(y_a, b) \wedge \mathsf{C}(z, y_a)]$ to $\exists b[\mathsf{P}(x_a, b) \wedge \mathsf{C}(z, b)$. This is proved as follows: Let be given $\exists y_a \in (a)[\mathsf{P}(x_a, b) \wedge$

$\mathsf{P}(y_a, b) \wedge \mathsf{C}(z, y_a)]$. As $\mathsf{C}(z, y_a)$ and $\mathsf{P}(y_a, b)$ hold, one also has $\mathsf{C}(z, b)$. For the other direction one can choose for y_a just region x_a.

The *basic spatial relatedness relation rcc-sr* between an arbitrary region z and a cell x_a from a partition chain pc containing partition (a) is defined as follows:

$$\forall x \in (a) \forall z[\text{rcc-sr}(z, x_a) \text{ iff there is } b \in (a)^{\uparrow}[\mathsf{P}(x_a, b) \wedge \mathsf{C}(z, b)]] \tag{3.5}$$

This definition can be easily extended to arbitrary regions as the second argument and leads to the official definition of RCC spatial relatedness rcc-sr given by the condition below, which is a straight forward adaptation of the original definition.

Definition 3.36 The RCC spatial relatedness relation rcc-sr between arbitrary regions z and x w.r.t. a partition chain pc containing partition (a) is defined as follows:

$$\begin{aligned} \text{rcc-sr}(z, x) \text{ iff } \quad & \text{for } \tilde{x} = \text{cell on lowest level s.t. } \mathsf{P}(x, \tilde{x}) \\ & \text{there is } b \in (a)^{\uparrow}[\mathsf{P}(\tilde{x}, b) \wedge \mathsf{C}(z, b)] \end{aligned} \tag{3.6}$$

So, a region z is RCC spatially related to a region x iff z is spatially connected to the upper cell b of the smallest cell $\tilde{x}$ containing x. In other words, z may stand in one of the seven basic RCC8 relations different from dc to b.

A special case of RCC spatial relatedness is the one where z and x both are cells (not necessarily from the same level), as this excludes the basic relation po between z and b (due to the disjointness conditions for cells in a partition chain). So, using the fact that $\mathsf{P} = \{\mathsf{eq}, \mathsf{tpp}, \mathsf{ntpp}\}$, the following simple observation can be made.

Proposition 3.37 *For all cells z and cell x_a from partition (a) in partition chain pc, the following equivalence holds:*

$$\textit{rcc-sr}(z, x_a) \text{ iff } \exists b \in (a)^{\uparrow}[\mathsf{P}(x_a, b) \wedge (\mathsf{P}(b, z) \vee \mathsf{P}(z, b) \vee \mathsf{ec}(z, b))]]$$

As a consequence, in order to determine spatial relatedness of two cells—without the need of complex reformulations—it suffices that the data provide containment relations (part-of relation P) and neighbourhood relations (externally connected relation ec) between cells. And indeed, in general, most of the geographical linked open data provide these basic relations up to some degree of completeness (recall) and correctness (precision) as shown in [11].

Example 3.38 Consider again the partition chain of Fig. 3.1. The only ld-regions x_a that are spatially related to region c_2, formally: rcc-sr(x_a, c_2), are $x_a = c_1$, $x_a = c_2$, $x_a = c_4$, $x_a = b_1$ and $x_a = b_2$.

Let *pc* be a normal partition chain and $\mathcal{A} = \mathcal{A}_{pc}$ be its representation by predicate logical axioms. Moreover, let $KB = \mathcal{A} \cup Ax_{\text{BRCC}} \cup \{(3.5)\}$ be the knowledge base representing the partition chain plus the axioms for the boolean region connection calculus plus the definition of basic spatial relatedness (3.5). The investigations of the logical properties of the nearness relations are done with respect to this knowledge base *KB*.

Proposition 3.39 *For all cells* $x_a \in (a_i)_{i \in I}$, $KB \models rcc\text{-}sr(x_a, x_a)$.

The relation rcc-sr is not symmetric and not transitive in the general case. This can be explained by the fact that it is the second argument which determines the comparison context. However, if x_a, y_a are neighboring regions of the same partition (a), then symmetry (but not transitivity) holds. Accordingly, rcc-sr is called a "weakly asymmetrical" relation in [12].

Proposition 3.40 *For all* $x_a, y_a \in (a)$*, if* $KB \models \mathsf{EC}(y_a, x_a)$*, then* $KB \models rcc\text{-}sr(y_a, x_a)$ *and* $KB \models rcc\text{-}sr(x_a, y_a)$.

Proof See p. 71. □

The question which RCC8 base relations r are sufficient for spatial relatedness, i.e., for which $r \in \mathcal{B}_{RCC8}$ does $r(z, x)$ entail rcc-sr(z, x), is answered in Proposition 3.41.

Proposition 3.41 *For all* z, x_a:
If $KB \models z\{\mathsf{EC}, \mathsf{PO}, \mathsf{EQ}, \mathsf{TPP}, \mathsf{NTPP}, \mathsf{TPPi}, \mathsf{NTPPi}\}x_a$*, then* $KB \models rcc\text{-}sr(z, x_a)$.

This follows directly from (3.5) for $b = x_a$. As a corollary of this proposition and the definition of rcc-sr one can see that all cells are in rcc-sr-relation to cells (of upper levels) of which they are a part. Similarly, all cells are in rcc-sr-relation to cells (of lower levels) which they contain.

Relation rcc-sr(z, x_a) is independent of all base relations of RCC8 in the following sense: one can find for any base relation $r \in \mathcal{B}_{RCC8}$ regions z and x_a such that $KB \models$ rcc-sr$(z, x_a) \wedge r(z, x_a)$. Hence, if one knows that z is spatially related to x_a one cannot infer anything about the RCC8 base relation holding between them. Particularly, one cannot infer that z and x_a must be connected.

Regarding the properties related to proximities similar observations as that for non-RCC spatial relatedness can be proved.

Note again that now the proximity definitions are adapted to the RCC scenario where $\cup$ stands for summation and non-empty intersection between regions a and b amounts to saying that not $\mathsf{DC}(a, b)$ (or equivalently: $\mathsf{C}(a, b)$).

Proposition 3.42 *Let* $(a) = (a_i)_{i \in I}$ *be a partition consisting of cells. For all regions* $A \subseteq X$ *and all* $B \in (a)^{\uparrow}$ *with* $B = b_1 \cup \ldots \cup b_n$ *for* $b_j \in (a)$ *and* $j \in \{1, \ldots, n\}$ *the following entailment holds: if* $KB \models rcc\text{-}sr(A, b_1)$ *or … or* $KB \models rcc\text{-}sr(A, b_n)$*, then* $KB \models rcc\text{-}sr(A, B)$.

Proof See p. 72. □

That the other direction does not hold can be illustrated with the previous Example 3.4 on p. 52.

Of course for the left component the additivity condition holds as verified by the following proposition:

Proposition 3.43 *For all* $A, B, C \subseteq X$, $KB \models rcc\text{-}sr(A \cup B, C)$ *iff* $KB \models rcc\text{-}sr(A, C)$ *or* $KB \models rcc\text{-}sr(B, C)$.

Proof See p. 72. □

Adapting the notion of irregular split to the RCC scenario, in the end the following proposition results:

Proposition 3.44 *rcc-sr is a regular right-scaled proximity relation.*

Proof See p. 72. □

Moreover one can show that the relation rcc-sr based on a normal partition chain fulfils the connecting property, i.e., every region is near its complement or vice versa.

Proposition 3.45 *Let be given a nearness relation rcc-sr based on a normal partition chain pc according to the definition in 3.5. Then for all* $A \subseteq X$ *it holds that* $rcc\text{-}sr(A, X \setminus A)$ *or* $rcc\text{-}sr(X \setminus A, A)$.

Proof See p. 73. □

The conditions stated in a right-scaled proximity space are not strong enough to define a canonical topological space as is done for proximity spaces (see above). But, nonetheless, spatial relatedness can be seen as an interleaving of level-fixed relations. This will be explicated in the following. Let be given a total ordering of partitions $(a_i^j)_{i \in I_j}$ over $X, 1 \leq j \leq n$. For every partition level j, define spatial relatedness $\widetilde{\text{rcc-sr}}^j$ between arbitrary regions $z_1, z_2 \subseteq X$ as follows:

$$\text{rcc-sr}^j(z_1, z_2) \text{ iff there is a } y \text{ of level } j \text{ s.t. rcc-sr}(z_1, y) \text{ and rcc-sr}(z_2, y).$$

These relations are symmetric and are ordered with respect to inclusion and fulfil the conditions of a minimal proximity space.

Proposition 3.46 *The level-fixed nearness relations* $rcc\text{-}sr^j$ *fulfil the following conditions:*

1. *Every* $rcc\text{-}sr^j$ *is a (symmetric) proximity relation.*
2. *If* $i \leq j$, *then* $rcc\text{-}sr^i \subseteq rcc\text{-}sr^j$.
3. *If* $rcc\text{-}sr(z_1, z_2)$ *and* z_2 *is of level* j, *then* $rcc\text{-}sr^j(z_1, z_2)$.

Proof See p. 73. □

As a conclusion it can be stated that though rcc-sr is not a (minimal) proximity relation each of its levels induces a proximity relation rcc-sr^j extending rcc-sr.

An open problem is a representation theorem for RCC-based spatial relatedness. Prima facie, it is not possible to adapt the proof of the representation theorem for non-RCC based spatial relatedness as there I used the finiteness of X and argued with singletons $\{x\} \subseteq X$.

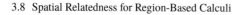

3.9 Related Work

The general idea of a scaling context for spatial relations (more specifically: nearness relations) goes back to the work of [27]. But in contrast to the work of this chapter, the approaches outlined in [27] as well as in the following work [3, 16, 28] do not deal with axiomatic characterisations.

The definition of spatial relatedness in this monograph follows a general "information processing" strategy that can be found in different areas of computer science. For example, belief revision [1] (see also Chap. 7) is concerned with the general task of integrating a new piece of information a into a knowledge base b. If a is not compatible (associate: spatially related) to b, then one weakens b to a set b' with $b \models b'$ and $b' \cup \{a\} \not\models \bot$ by throwing out elements from b. The KB b' is less strict and thus less informative than b. This can be understood as making b' more similar to X.

A similar situation occurs in the sub-area of knowledge-based reasoning called abduction where one has to find explanations for observations [15, 17]. In most cases, observations cannot be deduced from the theory or facts at hand, but have to be found within a space of possible explanations. The idea is to keep the creativity effort needed as low as possible—going only a minimal step upwards in the explanation space.

The underlying structure of sr_{pc} are partition chains, which are special trees. The work of [25] and [13] focus on the dynamics of such tree structures, called adjacency trees. In contrast, this monograph uses these tree structures as a basis for a spatial relatedness definition and gives a full axiomatic characterisation.

3.10 Summary

This chapter gave a fine-grained semantical analysis of spatial relatedness based on partition chains—resulting in a representation theorem for the special case where the partition chain is strict and where the arguments are subsets of a finite set X. Users of information systems such as agents that move, act, and plan in an environment according to some internal qualitative spatial map or semantic web systems [2] relying on spatio-temporal ontologies may benefit from the representation result given in Sect. 3.6 because it completely characterises the spatial relations at hand.

More concretely, if an agent or a query answering system would rely only on axiomatic characterisations of a spatial relatedness relation, then it would have to incorporate a deduction engine in order to do planning or query answering. The reason is that only if the agent considers all entailments of the axioms for spatial relatedness relations, it can be guaranteed that he will reach all possible plan configurations or all possible answers, respectively. On the other hand, knowing that the axioms for spatial relatedness have exactly one model (modulo renaming of the domain elements), the agent/the system may directly work with the model and apply, e.g., model checking—which is usually more feasible regarding complexity than calculating the deductive closure of a set of axioms.

A glimpse of how the dynamics of partition chains affect spatial relatedness was given in Sect. 3.7. The work described in that section gives rise to further interesting research questions and tasks. The simplest one is the task of investigating the effects of merging for levels below the next-to-last level. In this case one will have to distinguish between merging regions with the same upper level cells vs. merging regions with different upper level cells. Additionally one has to define how to propagate the merge effect to the higher levels (as the merger on level i may affect also cells on levels above $i + 1$.) Moreover, next to the merge operation adaptations of other change operations mentioned by [14] to the partition-chain framework could be investigated.

Regarding the coverage of applications, the spatial-relatedness framework applies only to constellations with a total order of nested partitions. In order to cover other applications, one would have to give up the condition on nestedness—thereby making the induced spatial relations more interesting but even harder to characterise. Typical examples for non-nested regions are so-called micro functional regions [10].

The extension of the spatial relatedness to regions in the RCC calculus is straight forward. But as mentioned before a representation theorem for spatial relatedness on RCC regions still has to be given.

References

1. Alchourrón, C.E., Gärdenfors, P., Makinson, D.: On the logic of theory change: Partial meet contraction and revision functions. Journal of Symbolic Logic **50**, 510–530 (1985)
2. Berners-Lee, T., Hendler, J., Lassila, O.: The semantic web. Scientific American **284**(5), 34–43 (2001)
3. Brennan, J., Martin, E.: Spatial proximity is more than just a distance measure. International Journal of Human-Computer Studies **70**, 88–106 (2012)
4. Cohn, A.G., Bennett, B., Gooday, J., Gotts, N.M.: Qualitative Spatial Representation and Reasoning with the Region Connection Calculus. Geoinformatica **1**, 275–316 (1997)
5. Dimov, G., Vakarelov, D.: Contact algebras and region-based theory of space: A proximity approach - I. Fundamentae Informaticae **74**(2-3), 209–249 (2006)
6. Dimov, G., Vakarelov, D.: Contact algebras and region-based theory of space: Proximity approach - II. Fundamentae Informaticae **74**(2-3), 251–282 (2006)
7. Düntsch, I., Vakarelov, D.: Region-based theory of discrete spaces: A proximity approach. Annals of Mathematics and Artificial Intelligence **49**, 5–14 (2007)
8. Efremovič, V.A.: Infinitesimal spaces. Dokl. Akad. Nauk SSSR **76**, 341–343 (1951)
9. Efremovič, V.A.: The geometry of proximity I. Mat. Sb. (N.S.) **31**(73), 189–200 (1952)
10. Grütter, R., Helming, I., Speich, S., Bernstein, A.: Rewriting queries for web searches that use local expressions. In: N. Bassiliades, G. Governatori, A. Paschke (eds.) Proceedings of the 5th International Symposium on Rule-Based Reasoning, Programming, and Applications (RuleML-11), *LNCS*, vol. 6826, pp. 345–359 (2011)
11. Grütter, R., Purves, R.S., Wotruba, L.: Evaluating topological queries in linked data using dbpedia and geonames in switzerland and scotland. Transactions in GIS **21**(1), 114–133 (2017). DOI 10.1111/tgis.12196
12. Grütter, R., Scharrenbach, T., Waldvogel, B.: Vague spatio-thematic query processing: A qualitative approach to spatial closeness. Transactions in GIS **14**(2), 97–109 (2010)
13. Jiang, J., Worboys, M.: Event-based topology for dynamic planar areal objects. International Journal of Geographical Information Science **23**(1), 33–60 (2009)

14. Kauppinen, T., Väätäinen, J., Hyvönen, E.: Creating and using geospatial ontology time series in a semantic cultural heritage portal. In: Proceedings of the 5th European Semantic Web Conference on the Semantic Web (ESWC-08): Research and Applications, pp. 110–123. Springer-Verlag, Berlin, Heidelberg (2008)
15. Kowalski, R.: Computational Logic and Human Thinking: How to be Artificially Intelligent. Cambridge University Press, New York (2011)
16. Mata, F.: Geographic Information Retrieval by Topological, Geographical, and Conceptual Matching. In: F. Fonseca, M.A. Rodríguez, S. Levashkin (eds.) Proceedings of the 2nd International Conference on GeoSpatial Semantics (GeoS-07), *LNCS*, vol. 4853, pp. 98–113. Springer, Berlin Heidelberg (2007)
17. Möller, R., Özçep, Ö., Haarslev, V., Nafissi, A., Wessel, M.: Abductive conjunctive query answering w.r.t. ontologies. KI - Künstliche Intelligenz pp. 1–6 (2015). DOI 10.1007/s13218-015-0399-3
18. Naimpally, S., Warrack, B.D.: Proximity Spaces. No. 59 in Cambridge Tracts in Mathematics and Mathematical Physics. Cambridge University Press (1970)
19. Özçep, Ö.L., Grütter, R., Möller, R.: Nearness rules and scaled proximity. In: L.D. Raedt, C. Bessiere, D. Dubois (eds.) Proceedings of the 20th European Conference on Artificial Intelligence (ECAI-12), pp. 636–641 (2012)
20. Özçep, Ö.L.: A representation theorem for spatial relations. In: B. Pfahringer, J. Renz (eds.) Proceedings of the 28th Australasian Joint Conference on Artificial Intelligence 2015 (AI-15), *LNAI*, vol. 9457, pp. 444–456 (2015)
21. Özçep, Ö.L., Grütter, R., Möller, R.: Dynamics of a nearness relation–first results. In: Proceedings of the International Workshop on Spatio-Temporal Dynamics (SteDy-12) (2012)
22. Randell, D.A., Cui, Z., Cohn, A.G.: A spatial logic based on regions and connection. In: Proceedings of the 3rd International Conference on Knowledge Representation and Reasoning (KR-92), pp. 165–176 (1992)
23. Riesz, F.: Stetigkeitsbegriff und abstrakte Mengenlehre. In: Atti del IV Congresso Intern. dei Matem., Bologna, vol. 2, pp. 18–24 (1908)
24. Smyth, M.B., Webster, J.: Discrete spatial models. In: M. Aiello, I. Pratt-Hartmann, J. Benthem (eds.) Handbook of Spatial Logics, pp. 713–798. Springer Netherlands (2007)
25. Stell, J.G., Worboys, M.F.: Relations between adjacency trees. Theoretical Computer Science **412**(34), 4452–4468 (2011)
26. Vakarelov, D., Dimov, G., Düntsch, I., Bennett, B.: A proximity approach to some region-based theories of space. Journal of Applied Non-Classical Logics pp. 527–559 (2001)
27. Worboys, M.F.: Nearness relations in environmental space. International Journal of Geographical Information Science **15**(7), 633–651 (2001)
28. Yao, X., Thill, J.C.: How Far Is Too Far? – A Statistical Approach to Context-contingent Proximity Modeling. Transactions in GIS **9**(2), 157–178 (2005)

Appendix

Proof of Proposition 3.8

Ad (P1): By definition. Ad (P2): Holds because the upper-shift of $b \cup c$ contains the upper shifts of b and c. Ad (P3): Left-to-right clear. For the other direction: As $a \cup b$ has a non-empty intersection with the upper-shift of c, one of a or b must have a non-empty intersection with the upper-shift of c, hence either $\mathrm{sr}_{pc}(a, c)$ or $\mathrm{sr}_{pc}(b, c)$. Ad (P4): If $a \cap b \neq \emptyset$, then in particular both a and b are non-empty. The assertion follows because the upper-shift of b contains b. Ad (P5): Follows from the fact that there must be a proper set extension with an upshift of either a or $X \setminus a$.

Proof of Proposition 3.12

Clearly one has $(b)^{\Uparrow,pc} \subseteq (b)^{\Uparrow_{\mathrm{sr}pc}}$: $(b)^{\Uparrow,pc}$ contains all cells b' on the same level as b for which $b \overset{\bullet}{\sim} b'$, hence $b' \subseteq (b)^{\Uparrow_{\mathrm{sr}pc}}$. For the other direction assume for contradiction that there is some a with $b \overset{\bullet}{\sim} a$ and $a \not\subseteq (b)^{\Uparrow,pc}$. Let $a' := a \setminus (b)^{\Uparrow,pc}$. It holds that $\mathrm{sr}(a', a)$, as $a' \cap a \neq \emptyset$. Because of $b \overset{\bullet}{\sim} a$ it follows that also $\mathrm{sr}(a', b)$ contradicting $a' \cap (b)^{\Uparrow,pc} = \emptyset$.

Proof of Proposition 3.14

1. Let $\delta(a, c)$ or $\delta(b, c)$. Because of (Pgreln) it follows that $a \cap c^{\Uparrow_\delta} \neq \emptyset$ or $b \cap c^{\Uparrow_\delta} \neq \emptyset$. Hence $(a \cup b) \cap c^{\Uparrow_\delta} \neq \emptyset$ and with (Pgrels) it follows that $\delta(a \cup b, c)$. The other direction is proved similarly: Assume $\delta(a \cup b, c)$, then because of (Pgreln) it follows $(a \cup b) \cap c^{\Uparrow_\delta} \neq \emptyset$, i.e., $a \cap c^{\Uparrow_\delta} \neq \emptyset$ or $b \cap c^{\Uparrow_\delta} \neq \emptyset$ which with (Pgrels) entails $\delta(a, c)$ or $\delta(b, c)$.
2. Let $a \cap b \neq \emptyset$. In particular $b \neq \emptyset$. It is the case that $b \overset{\bullet}{\sim} b$, hence $b \subseteq b^{\Uparrow_\delta}$. Hence $a \cap b^{\Uparrow_\delta} \neq \emptyset$, so with (Pgrels) it follows that $\delta(a, b)$. The same argument works for the roles of a and b exchanged.

Proof of Proposition 3.15

Let $a \cap b^{\Uparrow_\delta} \neq \emptyset$. Then there is a c such that $c \overset{\bullet}{\sim} b$ and $a \cap c \neq \emptyset$. Because of (P4) one gets $\delta(a, c)$. So due to the definition of $\overset{\bullet}{\sim}$ one must also have $\delta(a, b)$.

Proof of Proposition 3.16

Let $a \subseteq b$. Let $x \in a^{\Uparrow_\delta}$. So there is a c with $c \overset{\bullet}{\sim} a$ and $x \in c$. I show $c \subseteq b^{\Uparrow_\delta}$ (and so also $x \in b^{\Uparrow_\delta}$). Assume not, then there is a subset of $c' \subset c$ such that $c' \cap b^{\Uparrow_\delta} = \emptyset$. Because of (Pgreln) it follows that not $\delta(c', b)$. Because of (P2) also not $\delta(c', a)$, and because of $c \overset{\bullet}{\sim} a$ also not $\delta(c', c)$, but this contradicts (P4).

Proof of Proposition 3.17

Due to Proposition 3.12 one can assume that $\cdot^{\Uparrow_\delta} = (\cdot)^{\Uparrow^{\mathrm{sr}pc}} = (\cdot)^{\Uparrow,pc}$. Now, let $a^{\Uparrow_\delta} \subsetneq b^{\Uparrow_\delta}$, so the pc-cell $a^{\Uparrow_\delta}$ is a proper subset of the cell $b^{\Uparrow_\delta}$. But then an additional application of $\cdot^{\Uparrow_\delta} = (\cdot)^{\Uparrow,pc}$ amounts to a shift in pc which must be a cell contained in or the same as the cell $b^{\Uparrow_\delta}$. This shows (Pdoubleshift).

In order to prove (Punionshift), let ${a^{\Uparrow_\delta}}^{\Uparrow_\delta} = {b^{\Uparrow_\delta}}^{\Uparrow_\delta} \neq X$ and $a^{\Uparrow_\delta} \not\subseteq b^{\Uparrow_\delta}$ and $b^{\Uparrow_\delta} \not\subseteq a^{\Uparrow_\delta}$. So the pc-cells $a^{\Uparrow_\delta}$ and $b^{\Uparrow_\delta}$ are not on the same path from the leaves to the root X in pc, but their upper shifts are. So, that means that $\widetilde{a \cup b}$ must be a cell that contains $a^{\Uparrow_\delta}$ and $b^{\Uparrow_\delta}$. As it is the smallest such cell one gets $\widetilde{a \cup b} = {a^{\Uparrow_\delta}}^{\Uparrow_\delta} = {b^{\Uparrow_\delta}}^{\Uparrow_\delta}$. Hence $a \cup b^{\Uparrow_\delta} = {{a^{\Uparrow_\delta}}^{\Uparrow_\delta}}^{\Uparrow_\delta} = {{b^{\Uparrow_\delta}}^{\Uparrow_\delta}}^{\Uparrow_\delta}$.

Proof of Proposition 3.18

Let $a^{\Uparrow_\delta} \not\subseteq b^{\Uparrow_\delta}$ and $b^{\Uparrow_\delta} \not\subseteq a^{\Uparrow_\delta}$. So the pc-cells $a^{\Uparrow_\delta}$ and $b^{\Uparrow_\delta}$ are not on the same path from the leaves to the root X in pc. That means that $\widetilde{a \cup b}$ must be a cell whose underlying set contains $a^{\Uparrow_\delta}$ and $b^{\Uparrow_\delta}$. Now consider $(a' \cup b')$. I prove $\mathrm{us}((\widetilde{a' \cup b'})^{\uparrow,pc}) = \mathrm{us}((\widetilde{a \cup b})^{\uparrow,pc})$. As $a' \cup b' \subseteq a \cup b$, it follows that $\mathrm{us}(\widetilde{a' \cup b'}) \subseteq \mathrm{us}(\widetilde{a \cup b})$, so one knows that the cell $\widetilde{a' \cup b'}$ must be under the cell $\widetilde{a \cup b}$. Now assume that $\widetilde{a \cup b}$ is strictly above the cell $\widetilde{a' \cup b'}$. As $a^{\Uparrow_\delta}$ and $b^{\Uparrow_\delta}$ are incomparable it must be the case that $\mathrm{us}(\widetilde{a' \cup b'}) \supseteq a^{\Uparrow_\delta} \cup b^{\Uparrow_\delta}$. Hence $\mathrm{us}(\widetilde{a' \cup b'}) \supseteq \mathrm{us}(\widetilde{a^{\Uparrow_\delta} \cup b^{\Uparrow_\delta}}) \supseteq \mathrm{us}(\widetilde{a \cup b})$ which means that the underlying set of $\widetilde{a \cup b}$ is contained in the underlying set of the cell $\widetilde{a' \cup b'}$. With the assertion proven before this would mean that $\widetilde{a \cup b}$ and $\widetilde{a' \cup b'}$ have the same underlying sets. Could it be the case that the level of $\widetilde{a \cup b}$ is strictly higher than that of $\widetilde{a' \cup b'}$? No, because $\widetilde{a \cup b}$ the smallest low level cell containing $a \cup b$ and this must be the level of $\widetilde{a' \cup b'}$ as the underlying set is the same as that of $\widetilde{a \cup b}$.

Proof of Proposition 3.21

Assume $\mathsf{uiso}(a)$. In order to show $\mathsf{siso}(a)$, let $\delta(x, a)$, then (Pgreln) says that $x \cap a^{\Uparrow_\delta} \neq \emptyset$, but $a^{\Uparrow_\delta} = a$, so $x \cap a \neq \emptyset$. Now assume $\mathsf{siso}(a)$. It has to be shown that $a^{\Uparrow_\delta} = a$, indeed, it is sufficient to show $a^{\Uparrow_\delta} \subseteq a$. So let $b \sim^{\bullet} a$, that means that all b have the same set of incoming δ edges as a. I have to show $b \subseteq a$. Assume otherwise, let $e = b \setminus a$. It holds that $\delta(e, b)$. So it must also hold that $\delta(e, a)$. But as siso, this entails $e \cap a \neq$, contradiction.

Proof of Proposition 3.23

Assume $\mathrm{sr}(a, b \uplus c)$ and not $\mathrm{sr}(a, b)$ and not $\mathrm{sr}(a, c)$. As $\mathrm{us}(b^{\Uparrow}) \cap a = \emptyset$ and $\mathrm{us}((c^{\Uparrow}) \cap a = \emptyset$ but $\mathrm{us}((b \uplus c)^{\Uparrow}) \cap a \neq \emptyset$, one has $\mathrm{us}(b^{\Uparrow}) \cup \mathrm{us}(c^{\Uparrow}) \subsetneq \mathrm{us}((b \uplus c)^{\Uparrow})$. Hence it follows that $\mathrm{us}((b^{\Uparrow}) \neq \mathrm{us}(c^{\Uparrow})$, because otherwise one would have $\mathrm{us}(b^{\Uparrow}) \cup \mathrm{us}(c^{\Uparrow}) =$

$\mathrm{us}((b \uplus c)^{\Uparrow})$. Now, let $b \uplus c = b' \uplus c'$ where $b' \neq b$ and $c \neq c'$. One of b', c' must have elements of both b and c. W.l.o.g let us assume it is b'. That means that $\widetilde{b'} = \widetilde{b \cup c}$ and hence $\mathrm{sr}(a, b')$.

Proof of Proposition 3.25

Clearly reflexivity and symmetry hold for $\sim_0$. So it remains to show transitivity. So let $x \sim_0 y$ and $y \sim_0 z$. One has to show that $x \sim_0 z$. For contradiction assume $x \nsim_0 z$. Then either not $\{x\} \overset{\bullet}{\sim} \{x, z\}$ or not $\{z\} \overset{\bullet}{\sim} \{x, z\}$. Assume it is not $\{x\} \overset{\bullet}{\sim} \{x, z\}$. (The other case is handled symmetrically.) As (P2) holds, this can be the case only if there is a set a such that $\delta(a, \{x, z\})$ but not $\delta(a, \{x\})$. The latter together with the assumptions that $x \sim_0 y$ and $y \sim_0 z$ implies, that not $\delta(a, \{x, y\})$ and not $\delta(a, \{y\})$ and not $\delta(a, \{y, z\})$ and not $\delta(a, \{z\})$. But now, as $\delta(a, \{x, z\})$, axiom $(P2)$ implies also $\delta(a, \{x, z, y\})$. But this means that one has two different irregular splits of $\{x, z, y\}$ w.r.t. a, namely $\{x\} \uplus \{z, y\}$ and $\{z\} \uplus \{x, y\}$. This contradicts Axiom (PirrSplit), hence one may conclude that $\sim_0$ is indeed an equivalence relation and that for all $x \in X$, the equivalence class $[x]_{\sim_0}$ is defined.

Proof of Theorem 3.26

Let $b_1, b_2 \subseteq [x]_{\sim_0}$. If b_1, b_2 are singletons, then the assertion follows directly from the assumption. Now assume that one of them contains two elements, say it is $b_1 = \{e, f\}$ and $b_2 = \{d\}$. One knows

$$b_1 = \{e, f\} \overset{\bullet}{\sim} \{f\} \overset{\bullet}{\sim} \{f, d\} \overset{\bullet}{\sim} \{d\} = b_2$$

If $b_2 = \{d_1, d_2\}$ then one has

$$b_1 = \{e, f\} \overset{\bullet}{\sim} \{f\} \overset{\bullet}{\sim} \{f, d_1\} \overset{\bullet}{\sim} \{d_1\} \overset{\bullet}{\sim} \{d_1, d_2\} = b_2$$

Now a more restricted version of the proposition is proved: namely, for all $b \subseteq [x]_{\sim_0}$ one has for all $x \in b$: $x \overset{\bullet}{\sim} b$. This is true for b of size 1 and 2. So assume for induction that it holds for all b of size n and assume that b has size $n + 1$, e.g., $b = b' \cup \{e\}$. Take an arbitrary $z \in b' \cup \{e\}$. In the first case say $z \in b'$. By induction: $z \overset{\bullet}{\sim} b'$ and $z \overset{\bullet}{\sim} e$. If not $z \overset{\bullet}{\sim} b' \cup \{e\}$ were the case, then this could only be the case because there exists f with $\delta(f, b' \cup \{e\})$ but not $\delta(f, z)$, that would also mean that not $\delta(f, b')$ and not $\delta(f, e)$. So one gets an irregular split of $b' \cup \{e\}$ for f. As b' contains at least two elements, say $g \in b'$ and $g \neq e$, I consider now $b' \setminus \{g\} \cup \{e, g\}$. Now it must be not $\delta(f, b' \setminus \{g\})$ and not $\delta(f, \{e, g\})$ as $e \overset{\bullet}{\sim} \{e, g\}$. So in the end, more than one irregular splitting exists, contradiction. Now of course, take b_1 and b_2 arbitrarily. Then $b_1 \overset{\bullet}{\sim} x$ for any $x \in b_1$ and $b_2 \overset{\bullet}{\sim} y$ for any $y \in b_2$, but $x \overset{\bullet}{\sim} y$, hence $b_1 \overset{\bullet}{\sim} b_2$.

Proof of Theorem 3.29

Cells of level 0 are constructed as $[x]_{\sim_0}$ which is possible due to Prop. 3.25. On top of these one constructs per recursion other partitions, showing that these indeed are partitions and that all cells built have the same rank.

Assume that one has already constructed cells up to level n, i.e., one has a partition of sets on level n and a corresponding equivalence relation $\sim_n$. In case the nth partition consists only of the set X, the construction is finished. Otherwise one defines the cells on level $n+1$ as sets of the forms $a^{\Uparrow\delta}$, where a is a cell on level n. I have to show that theses indeed make up a set partition. Clearly, these sets cover the whole set X: because there are no isolated points different from X due to (Pnoiso), one has $a \subsetneq a^{\Uparrow\delta}$. Due to (Pnested) one knows that the sets on level n+1 are going to be aligned. But this does not exclude that one set $a^{\Uparrow\delta}$ is a proper subset of another set $b^{\Uparrow\delta}$ for sets a, b on level n, i.e., assume for contradiction that $a^{\Uparrow\delta} \subsetneq b^{\Uparrow\delta}$. Due to (Pdoubleshift) it follows that $a^{\Uparrow\delta\Uparrow\delta} \subseteq b^{\Uparrow\delta}$. Now take any $x \in a$ and $y \in b$. I consider two cases: n = 0. So a, b are cells on level 0. Because of Thm. 3.26 one knows that $x^{\Uparrow\delta} = a^{\Uparrow\delta}$ and $y^{\Uparrow\delta} = b^{\Uparrow\delta}$. But then the following (in)equalities can be derived:

$$\begin{aligned} r(\{x\}) &\overset{\text{(Prop. 3.26)}}{=} r(a) \overset{\text{(Def. of } r(\cdot))}{=} 1 + r(a^{\Uparrow\delta}) \overset{\text{(Def. of } r(\cdot))}{=} 2 + r(a^{\Uparrow\delta\Uparrow\delta}) \\ &\overset{\text{(Pdoubleshift)}}{\geq} 2 + r(b^{\Uparrow\delta}) > 1 + r(b^{\Uparrow\delta}) \overset{\text{(Def. of } r(\cdot))}{=} r(b) \overset{\text{(Def. of } r(\cdot))}{=} r(y^{\Uparrow\delta}) \end{aligned}$$

So, one would get $r(\{x\}) > r(\{y\})$ contradicting (Phomrank). So, it holds that all cells of level 0 have the same rank.

If $n > 0$, then a and b have the forms $a = a'^{\Uparrow\delta}$ and $b = b'^{\Uparrow\delta}$ for cells a', b' on level $n-1$. One may assume that these have the same rank (per induction.) Similar as for case $n = 0$ one calculates the inequality $r(a') > r(b')$, which contradicts the induction hypothesis.

Now assume pc is the partition chain resulting from this construction. One has to show that $\delta = \mathrm{sr}_{pc}$. One has to show for all a, b that $\delta(a, b)$ iff $\mathrm{sr}_{pc}(a, b)$. Assume that b is a cell in pc. Then $\mathrm{us}(\tilde{b}) = b$ in pc, and $(b)^{\Uparrow,pc} = (\tilde{b})^{\uparrow,pc} = \mathrm{us}(\tilde{b})^{\Uparrow\delta} = b^{\Uparrow\delta}$. So with (Pgreln) and (Pgrels) one gets $\delta(a, b)$ iff $\mathrm{sr}_{pc}(a, b)$.

Now assume that b is an arbitrary set. Let $\tilde{b}$ be the cell in pc containing b. If one can show that $b^{\Uparrow\delta} = \mathrm{us}(\tilde{b})^{\Uparrow\delta}$, then one can use the same argument as above for the case $b = \mathrm{us}(\tilde{b})$. Now show by induction on the level N of $\tilde{b}$ that $b = \mathrm{us}(\tilde{b})$.

Case $N = 0$: Here $\tilde{b}$ is a cell on level zero. Because of Prop. 3.26 one knows that $b \,{}^{\bullet}\!\!\sim \mathrm{us}(\tilde{b})$, hence $b^{\Uparrow\delta} = \mathrm{us}(\tilde{b})^{\Uparrow\delta}$.

Case $N > 0$: Two sub-cases are distinguished, $N = 1$ and $N > 1$. Assume first that $N = 1$. So let $us(\tilde{b})$ have level 1. Then b is covered by a set $\{c_1, \ldots, c_k\}$ of cells of level 0. I argue that it cannot be the case that $b^{\Uparrow\delta} \subseteq \mathrm{us}(\tilde{b})$: take $x_1 \in c_1, x_2 \in c_2$. As x_1, x_2 are in different cells c_1, c_2 of level 0 one knows that not ($x_1 \,{}^{\bullet}\!\!\sim \{x_1, x_2\}$ and $x_2 \,{}^{\bullet}\!\!\sim \{x_1, x_2\}$). Assume without loss of generality that not $x_1 \,{}^{\bullet}\!\!\sim \{x_1, x_2\}$. That means that there is a z such that $\delta(z, \{x_1, x_2\})$ but not $\delta(z, \{x_1\})$. But one has $b^{\Uparrow\delta} \supseteq \{x_1, x_2\}^{\Uparrow\delta} \supseteq {x_1}^{\Uparrow\delta} = {c_1}^{\Uparrow\delta} = us(\tilde{b})$. Now it cannot be the case that $b^{\Uparrow\delta} = \mathrm{us}(\tilde{b})$, because then $\{x_1, x_2\}^{\Uparrow\delta} = \mathrm{us}(\tilde{b}) = {x_1}^{\Uparrow\delta}$. As $\delta(z, \{x_1, x_2\})$, this means by (Pgreln) that

$z \in \{x_1, x_2\}^{\Uparrow\delta} = x_1^{\Uparrow\delta}$. By (Pgrels) it then follows $\delta(z, x_1)$—contradiction. Hence, $b^{\Uparrow\delta} \subsetneq \mathrm{us}(\tilde{b})$. But still it could be the case that $b^{\Uparrow\delta} \subsetneq \mathrm{us}(\tilde{b})^{\Uparrow\delta}$. But then due to nestedness (Pnested) $b^{\Uparrow\delta}$ must contain properly at least one cell of level 0, say d. So one has $d^{\Uparrow\delta} \subsetneq b^{\Uparrow\delta}$, hence with (Pdoubleshift) it follows that $d^{\Uparrow\delta\Uparrow\delta} \subseteq b^{\Uparrow\delta} \subsetneq us\tilde{b}^{\Uparrow\delta}$. But as d is a cell of level 1, one must have $d^{\Uparrow\delta} = \mathrm{us}(\tilde{b})$ leading to a contradiction. Hence $b^{\Uparrow\delta} = \tilde{b}^{\Uparrow\delta}$ follows.

Case $N > 1$. Let $\tilde{b}$ be a cell of level $N > 1$ and let b be covered by a set $\{c_1, \ldots, c_k\}$ of cells of level $N - 1 > 0$. The cells c_1, c_2 can be represented as $c_1 = c_1'^{\Uparrow\delta}$ and $c_2 = c_2'^{\Uparrow\delta}$ for $N - 1$ level cells c_1', c_2'. One can choose c_1' and c_2' such that $b \cap c_i' \neq \emptyset$, i.e., there is $x_1 \in c_1' \cap B$ and $x_2 \in c_2' \cap B$. It follows that $c_1'^{\Uparrow\delta} \nsubseteq c_2'^{\Uparrow\delta}$ and $c_2'^{\Uparrow\delta} \nsubseteq c_1'^{\Uparrow\delta}$ and $c_1'^{\Uparrow\delta\Uparrow\delta} = c_2'^{\Uparrow\delta\Uparrow\delta} \neq X$. Using axiom (Punionshift) it follows that $c_1' \cup c_2'^{\Uparrow\delta} = c_1'^{\Uparrow\delta\Uparrow\delta\Uparrow\delta} = c_2'^{\Uparrow\delta\Uparrow\delta\Uparrow\delta} = \tilde{b}^{\Uparrow\delta}$. Because of (Pcelldet) it follows that $x_1 \cup x_2^{\Uparrow\delta} = \tilde{b}^{\Uparrow\delta}$ and so also $b^{\Uparrow\delta} = \tilde{b}^{\Uparrow\delta}$.

Proof of Proposition 3.31

1. This assertion follows from the fact, that for all $b \subseteq X$ with level at most $n-3$ (in pc_2) the upward cells in both pc_1 and pc_2 are identical, $b^{\Uparrow, pc_1} = b^{\Uparrow, pc_2}$. Hence, by definition of nearness it immediately follows that $\mathrm{sr}_{pc_1}(a, b)$ iff $\mathrm{sr}_{pc_2}(a, b)$.
2. In order to proof this assertion suppose $\mathrm{sr}_{pc_1}(a, b)$, i.e., $a \cap us(b^{\Uparrow, pc_1}) \neq \emptyset$. I distinguish different cases depending on the level $l_{pc_1}(b)$ of b in pc_1.
   Assume $l_{pc_1}(b) = n-2$, then $b^{\Uparrow, pc_1} = (n-1, c)$ for some set c on the level $n-1$. If $c = c_1$ or $c = c_2$, then $b^{\Uparrow, pc_2} = (n-1, c_1 \cup c_2)$. So from $a \cap us(b^{\Uparrow, pc_1}) \neq \emptyset$ one deduces $a \cap us(b^{\Uparrow, pc_2}) \neq \emptyset$, i.e. $\mathrm{sr}_{pc_2}(a, b)$. If c is an underlying set of another cell on level $n-1$, then one has $us(b^{\Uparrow, pc_2}) = (n-1, c)$ and hence also $\mathrm{sr}_{pc_2}(a, b)$.
   Now assume that $l_{pc_1}(b) = n - 1$. Then $\tilde{b}^{pc_1} = (n-1, c)$ for some set c on the partition level $n - 1$. Then one will have $\tilde{b}^{pc_2} = (n-1, c')$ for $c \subseteq c'$. Hence, $b^{\Uparrow, pc_2} = (n, X)$ and so $\mathrm{sr}_{pc_2}(a, b)$.
   Last assume that $l_{pc_1}(b) = n - 1$. In this case, the level of b in pc_2 may be $n - 1$ or n. But in any case, one has $b^{\Uparrow, pc_2} = (n, X)$, and therefore $\mathrm{sr}_{pc_2}(a, b)$.

Proof of Proposition 3.40

According to (3.5), if $KB \models$ rcc-sr(y_a, x_a), then $KB \models \mathsf{P}(x_a, b) \wedge \mathsf{C}(y_a, b)$. Since $KB \models \mathsf{EC}(x_a, y_a)$ either $\mathsf{P}(y_a, b)$ or $\mathsf{EC}(y_a, b)$ must hold, hence $\mathsf{C}(y_a, b)$.

Proof of Proposition 3.42

Assume that $KB \models$ rcc-sr(A, b_1) or ... or rcc-sr(A, b_n). It has to be shown that $KB \models$ rcc-sr(A, B). W.l.o.g assume that $KB \models$ rcc-sr(A, b_1) (otherwise rename b_i). This means that there is a $b \in (a)^{\uparrow}$ such that $KB \models \mathsf{P}(b_1, b) \wedge \mathsf{C}(A, b)$. Since, according to our assumption, b_1 is already contained in B, it follows that $b = B$. Thus, for $b_1 = B$ one has $KB \models \mathsf{P}(B, B) \wedge \mathsf{C}(A, B)$, hence $KB \models$ rcc-sr(A, B).

Proof of Proposition 3.43

"⇒": Assume $KB \models$ rcc-sr$(A \cup B, C)$, that is according to (3.5), $KB \models \mathsf{P}(C, y) \wedge \mathsf{C}(A \cup B, y)$ for an ld-region y of the next level above C. But then $KB \models \mathsf{C}(A, y)$ or $KB \models \mathsf{C}(B, y)$, which follows from the definition of $A \cup B$ as sum of A and B. For *reductio ad absurdum* assume that not $KB \models \mathsf{C}(A, y)$ and not $KB \models \mathsf{C}(B, y)$, which means that $KB \models \mathsf{DC}(A, y)$ and $KB \models \mathsf{DC}(B, y)$. We have seen that, according to the assumption, $KB \models \mathsf{C}(A \cup B, y)$. Hence, there is a w such that $\mathsf{P}(w, A \cup B)$ and $\mathsf{P}(w, y)$. The first implies $\mathsf{C}(w, A \cup B)$, hence $\mathsf{C}(w, A)$ or $\mathsf{C}(w, B)$, that is, $\mathsf{C}(A, w)$ or $\mathsf{C}(B, w)$. Together with $\mathsf{P}(w, y)$ it follows from the definition of the relation P (cf. Sect. 2.4) that $\mathsf{C}(A, y)$ or $\mathsf{C}(B, y)$, not $\mathsf{DC}(A, y)$ or not $\mathsf{DC}(B, y)$, contradiction. Finally, for $y = C$ one has $KB \models \mathsf{P}(C, C) \wedge \mathsf{C}(A, C)$ or $KB \models \mathsf{P}(C, C) \wedge \mathsf{C}(B, C)$, hence $KB \models$ rcc-sr(A, C) or $KB \models$ rcc-sr(B, C).
"⇐": W.l.o.g. assume $KB \models$ rcc-sr(A, C), that is according to (3.5), $KB \models \mathsf{P}(C, y) \wedge \mathsf{C}(A, y)$ for an ld-region y of the next level above C. From the axiom for the sum of regions (see Sect. 2.4) it follows that if $\mathsf{C}(A, y)$, then $\mathsf{C}(A \cup B, y)$ for an arbitrary region $B \subseteq X$, hence, together with $\mathsf{P}(C, y)$, $KB \models$ rcc-sr$(A \cup B, C)$.

Proof of Proposition 3.44

Assume $KB \models$ rcc-sr$(A, B \cup C)$ and not $KB \models$ rcc-sr(A, B) and not $KB \models$ rcc-sr(A, C). Let $\mathsf{P}(B, b)$ and $\mathsf{P}(C, c)$, for b, c ld-regions of the next level above B, C. Since not $\mathsf{C}(A, b)$ and not $\mathsf{C}(A, c)$, one has $b \cup c \subsetneq (B \cup C)^{\uparrow}$, where $(B \cup C)^{\uparrow}$ is an ld-region of the next upper level. It must be the case that $b \neq c$, otherwise not $KB \models$ rcc-sr$(A, B \cup C)$. Now, let $B' \cup C' = B \cup C$ where $B' \neq B$ and $C' \neq C$, that is, we move border points of B to C (or vice versa) in order to get a different irregular splitting $B' \cup C'$ of $B \cup C$ w.r.t. A; but B' and C' will not be regions (in the meaning of the word as used in this work) anymore. Hence, the uniqueness of irregular splits is conserved, as long as B and C are constrained to be regions.

Proof of Proposition 3.45

Let $A \subseteq X$ be an arbitrary region. One has to show rcc-sr$(A, X \setminus A)$ or rcc-sr$(X \setminus A, A)$. First assume that A or $X \setminus A$ are not ld-regions, e.g., w.l.o.g., assume A is not a ld-region. Then $\tilde{A}$ overlaps with $X \setminus A$ and we have rcc-sr$(X \setminus A, A)$ (cf. Propositions 3 and 8). Now assume that both A and $X \setminus A$ are ld-regions. But, because the order is normal, either $a \subsetneq a^{\uparrow}$ or $X \setminus A \subsetneq (X \setminus A)^{\uparrow}$, where $A^{\uparrow}$ and $(X \setminus A)^{\uparrow}$ are ld-regions of the next upper level. Hence, either rcc-sr$(X \setminus A, A)$ or rcc-sr$(A, X \setminus A)$.

Proof of Proposition 3.46

Ad 1: Symmetry follows from commutativity of the conjunction in (8). As I excluded the empty set as a region, condition (1) for proximity spaces is trivially fulfilled. Because of the symmetry of rcc-sr^j, it suffices to show that one of the conditions (2.a) or (2.b) is fulfilled. I show it for (2.b). Accordingly, rcc-sr$^j(A \cup B, C)$ iff there is a y of level j such that rcc-sr$(A \cup B, y)$ and rcc-sr(C, y). This is equivalent to saying there are y_1 and y_2 of level j such that $\mathsf{P}(y, y_1)$ and $\mathsf{C}(A \cup B, y_1)$ as well as $\mathsf{P}(y, y_2)$ and $\mathsf{C}(C, y_2)$. But *per definitionem* $\mathsf{C}(A \cup B, y_1)$ iff $\mathsf{C}(A, y_1)$ or $\mathsf{C}(B, y_1)$, so that the equivalence with rcc-sr$^j(A, C)$ or rcc-sr$^j(B, C)$ follows.
Ad 2: Let rcc-sr$^i(z_1, z_2)$, i.e., there is a y of level i s.t. rcc-sr(z_1, y) and rcc-sr(z_2, y). According to Definition 2, there is a y' of level j such that $\mathsf{P}(y, y')$. As rcc-sr is a right-scaled proximity relation, it follows that rcc-sr(z_1, y') and rcc-sr(z_2, y'), hence rcc-sr$^j(z_1, z_2)$.
Ad 3: Let rcc-sr(z_1, z_2) and let z_2 be of level j. Then for $y = z_2$ we have rcc-sr(z_1, y) and (because of reflexivity) rcc-sr(z_2, y), hence rcc-sr$^j(z_1, z_2)$.

Chapter 4
Scalable Spatio-Thematic Query Answering

Abstract Providing query answering facilities at the conceptual level of a geographic data model requires deduction, which is a demanding task within geographical information systems due to the size of the data that are stored in secondary memory. In particular, this is the case for deductive query answering w.r.t. spatio-thematic ontologies, which provide a logical conceptualisation of an application domain involving geographic data. Considering this challenging task, first-order logic (FOL) rewritability is a desirable feature for query answering over geo-thematic ontologies. Hence, there is a need for combined spatio-thematic logics that provide a sufficiently expressive query language allowing for FOL rewritability. This chapter reports on FOL rewritability results for various combined spatio-thematic logics. The first result is that a weak coupling of DL-Lite with the expressive region connection calculus RCC8 allows for FOL rewritability under a spatial completeness condition for the assertional component of the data, the so-called abox. Stronger couplings allowing for FOL rewritability are possible only for spatial calculi as weak as the low-resolution calculus RCC2. Already a strong combination of DL-Lite with the low-resolution calculus RCC3 does not allow for FOL rewritability.

4.1 Introduction

There is a need for reasoning over geographical data in almost any area in which geographical information systems (GIS systems for short) are used, e.g., damage classification for flooding scenarios, development of eco systems in forestry, or analysis of sociological and demoscopic aspects in urban areas—to mention just a few. But providing reasoning services over geographical data is a demanding task because of at least two reasons explained in the following on the basis of the TIGER/Line® GIS data of the US Census Bureau[1].

[1] `http://www.census.gov/geo/www/tiger/`

Ö. L. Özçep, *Representation Theorems in Computer Science*,
https://doi.org/10.1007/978-3-030-25785-9_4

The first main problem is to specify the concepts and relations of the geographical domain over which one wants to reason. The intended meanings are not given in a formal or logical language with a precise semantics but in most cases with some feature codes and explanations of the codes in natural language. This holds also for the TIGER/Line® GIS data which specify features like parks, rivers, hospitals with the MAF/TIGER Feature Class Code and describe the intended meanings in the manual [8]. Only basic subsumption relations, e.g. "All parks are governmental area" are directly modelled in the data. If one wanted to provide a consistency test that checks whether the intended semantics of the feature codes are indeed in accordance with the data, one would have to do the hard of work translating the natural language specifications in some formal language and then apply a theorem or tableau prover over the resulting set of axioms.

But even if one had success in translating the natural language descriptions into some formal language, it would not be guaranteed that the formal language—which would have to be expressive enough in order to capture the natural language descriptions[2]—is computationally feasible.

And here enters the second main problem for reasoning over GIS-data: the huge amount of geographical data, which are usually stored persistently in secondary memory and maintained with sophisticated indexing mechanisms, restricts the possibilities for expressive declarative knowledge representation and reasoning. It is common sense knowledge that higher representation capabilities lead to more complex reasoning services over the representation in terms of time and space resources. As mentioned in the introductory chapter, query answering over a knowledge base is by far more difficult than query answering over a pure database. The reason for the increased difficulty is that an ontology may have many different models, hence ontology-based query answering has to compute the answers w.r.t. all models according to the *certain-answer semantics* as explicated in Sect. 2.2. The need for efficient answering w.r.t. the certain-answer semantics is more than obvious in case of GIS knowledge bases where geographical data already consume space resources before any reasoning has started. For example, loading the TIGER/Line® shapefiles for the state New York in the relational database management system like SQLServer 2008 results in a database of roughly 7 GB.

Nonetheless, for some reasoning scenarios over GIS data it is possible to define sufficiently expressive logics that are computationally feasible. The idea of providing a conceptualisation over the data is to filter unintended models of the data. The more expressive the logic is the more unintended models can be filtered. But for some GIS scenarios it is not necessary to give such a complete conceptualisation, it suffices to filter out some smaller set of the unintended models. And so the idea is to make the logic for representing the knowledge only as much expressible as is necessary to exclude this smaller set.

I rely on the main idea of strict OBDA (see Sect. 2.2) of reducing reasoning services (here: satisfiability checking and query answering) over an ontology to model checking an FOL query over the data. As mentioned in the introductory chapter,

[2] For example take the explanation of Lake/Pond in the manual [8, Appendix, F-7] that reads "A standing body of water that is surrounded by land"

lightweight logics such as DL-Lite are tailored towards rewriting, so these are the starting logics for the envisioned spatio-thematic logic. Though the rewritten queries may become exponentially bigger than the original ones, there exist optimisations based on semantic indexes which encode entailed knowledge of the terminological part of the ontology [23]. So, FOL rewritability can mean a benefit—under optimisations. But it should be clear that these optimisations also need information regarding the abox. Hence, though the rewritings are independent of the abox, the optimisations are not.

DL-Lite per se [4] is not sufficient for use in scenarios of geographic information processing, as geographic scenarios require, among others, the representation of and deduction over spatial concepts. Hence, the work presented in this chapter investigates combinations of logics in the DL-Lite family with different members of the RCC family [19] (see Sect. 2.4 for the necessary notions) and answers the question whether FOL rewritability of DL-Lite is preserved in the combined logic.

FOL rewritability is quite a non-robust property because adding apparently harmless constructors to logics known to allow FOL rewriting can immediately lead to non-rewritability. This holds also for the combination of DL-Lite with the RCC calculi. In general, there are two apparent possibilities for preserving FOL rewritability in case one extends a logic (such as DL-Lite) that allows for FOL rewritability: 1. choosing a weakly expressive RCC calculus (RCC2 or RCC3) or 2. choosing a weak combination of DL-Lite with a possibly expressive RCC calculus such as RCC8. Regarding the second possibility, the weakness of the combination means that in the combined logic the construction of arbitrary RCC8 constraint networks in the intensional part (tbox) of the ontology is prohibited.

In this chapter, the following main positive and negative results w.r.t. FOL rewritability along the two combination strategies are presented: a weak combination of DL-Lite$^{\sqcap}_{\mathcal{F},\mathcal{R}}$ with RCC8 allows for FOL rewriting w.r.t. the query language GCQ^{+}. The query language GCQ^{+} extends UCQs with the possibility to refer to RCC networks. This can be proved by a perfect-rewriting algorithm and an adapted chase procedure. Furthermore, considering strong combinations of DL-Lite with the weaker RCC fragments RCC3 and RCC2, it can be proved that DL-Lite$^{\sqcap,+}_{\mathcal{F},\mathcal{R}}$(RCC3) does not allow for FOL rewriting of satisfiability checking whereas the weaker DL-Lite$^{\sqcap,+}_{\mathcal{F},\mathcal{R}}$(RCC2) does [15].

This chapter is structured as follows. Weak combinations of DL-Lite with the region connection calculus are described in Sect. 4.2. Section 4.3 gives an extended example for formalising an ontology in DL-Lite$^{\sqcap}_{\mathcal{F},\mathcal{R}}$(RCC8) and a query and illustrates the adapted perfect-rewriting algorithm. In Sect. 4.4, the last section before the sections on related work and the summary, I consider strong combinations of DL-Lite with weaker fragments of the region connection calculus.

The content of this chapter was published in the papers [15, 16, 17].

4.2 Weak Combinations of DL-Lite with RCC

In this section, I describe a weak coupling of DL-Lite with the most expressive region connection calculus fragment RCC8 [16]. In the next section I am going to explain its use(fulness) with an example GIS scenario. Presenting the example will give me the opportunity to introduce further concepts that are necessary to understand the discussions on stronger couplings of DL-Lite with the weaker region connection calculi RCC2 and RCC3.

The combination paradigm follows that of Lutz and Miličič [11] who combine $\mathcal{ALC}$ with the RCC8 and, more generally, with ω-admissible concrete domains [11, Def. 5, p. 7]. The combined logic $\mathcal{ALC}(RCC8)$ of [11] is well behaved in so far as testing concept subsumption is decidable. As I aim at FOL rewritability I have to be even more careful in choosing the right combination method. The main aim is to construct spatio-thematic description logics in which the combinations between the thematic abstract domain and the spatial domain are maintained by constructors available in the logic $\mathcal{ALC}$(RCC8) of [11]. The approach presented here diverges from the one of [11] in the point that I do not presuppose an ω-admissible domain but some finite set of FOL-sentences that express corresponding properties of ω-admissible domains. So, I explicitly represent the axioms of the domain rather than making calls to an oracle. The main reason for this shift from a concrete domain to a theory is the fact that it is simpler to use known techniques for query answering (e.g. the chase construction) with respect to some axioms than with respect to a concrete domain.

Formally, let *Rel* be a finite set of binary relation symbols, *Const* be a set of constants and T_ω be a finite set of sentences with respect to a signature containing *Rel* and *Const*. A network $\mathcal{N}$ is a set of sentences over $Rel \cup Const$ of the form $r_1(a^*, b^*) \vee \cdots \vee r_k(a^*, b^*)$ for $r_1, \ldots, r_k \in Rel$ and $a^*, b^* \in Const$. The network $\mathcal{N}$ is called *complete* if it contains only atomic sentences $r(a^*, b^*)$ and if, additionally, for all constants a^*, b^* in $\mathcal{N}$ there is a $r \in Rel$ such that $r(a^*, b^*) \in \mathcal{N}$. For two complete finite networks $\mathcal{N}, \mathcal{M}$ let $I_{\mathcal{N},\mathcal{M}}$ denote the atoms $r(a^*, b^*) \in \mathcal{N}$ such that a^*, b^* occur in both $\mathcal{N}$ and $\mathcal{M}$. The restriction $\mathcal{N}_{Const'}$ of a network to the set of constants $Const'$ is the subset of $\mathcal{N}$ restricted to those sentences containing only constants from $Const'$.

T_ω is an *ω-admissible theory* iff it fulfils the following conditions:

1. satisfiability: T_ω is satisfiable.
2. JEPD property: T_ω entails the JEPD property (jointly exhaustive and pairwise disjoint) for relations in *Rel*.
3. decidability: Testing whether a finite complete syntactic network $\mathcal{N}$ is satisfiable with respect to T_ω, i.e., testing whether $T_\omega \cup \mathcal{N}$ is satisfiable, is decidable.
4. patchwork property: if $\mathcal{N}, \mathcal{M}$ are finite complete networks that are satisfiable relative to T_ω, respectively, and if $I_{\mathcal{N},\mathcal{M}} = I_{\mathcal{M},\mathcal{N}}$, then $\mathcal{N} \cup \mathcal{M}$ is satisfiable relative to T_ω, too.
5. compactness: a complete network $\mathcal{N}$ is satisfiable relative to T_ω iff for every finite set of constants X occurring in $\mathcal{N}$ the restriction $\mathcal{N}_X$ is satisfiable relative

to T_ω.
(This property is trivially fulfilled by all FOL-theories because FOL has the compactness property.)

The ω-admissible theories that are considered in this chapter are the axiom sets Ax_{RCCi} for various RCC calculi RCCi (see Sect. 2.4).

I recapitulate the syntax and the semantics of the constructors of [11] that are used for the coupling of the thematic and the spatial domain. As this logic works with concrete domains (i.e., a fixed structure which is referred to), the DL signature σ, as discussed in Sect. 2.1.3, is extended with attributes. An *attribute* is a binary relation with first argument over the domain of interpretations and the second argument over the elements of the concrete domain. Moreover, in the logic of [11], one is allowed to build complex attributes, called *paths*, by building a chain of roles ending with an attribute. Here, I consider only paths of length at most 2.

A *path* U (of length at most 2) is defined as l for a fixed attribute l ("has location") or as $R \circ l$, the composition of the role symbol R with l. In this chapter, $R \circ l$ is abbreviated as $\tilde{R}$. The usual notion of an interpretation $\mathfrak{I}$ in this combined logic is slightly modified by using two separate domains $\Delta^{\mathfrak{I}}$ and $(\Delta^*)^{\mathfrak{I}}$. All symbols of the theory T_ω are interpreted relative to $(\Delta^*)^{\mathfrak{I}}$. Let r be an RCC-relation of some RCC-fragment. That is, let be given a set of base relations $\mathcal{B}_{RCCi}$ and $r = \{r_1, \ldots r_n\} \equiv r_1 \vee \cdots \vee r_n$ for $r_i \in \mathcal{B}_{RCCi}$. Then the interpretation function $\mathfrak{I}$ is constrained as follows:

- $l^{\mathfrak{I}} \subseteq \Delta^{\mathfrak{I}} \times (\Delta^*)^{\mathfrak{I}}$
- $r^{\mathfrak{I}} = r_1^{\mathfrak{I}} \cup \cdots \cup r_n^{\mathfrak{I}}$
- $(R \circ l)^{\mathfrak{I}} = \{(d, e^*) \in \Delta^{\mathfrak{I}} \times (\Delta^*)^{\mathfrak{I}} \mid \text{ there is an } e \text{ s.t. } (d, e) \in R^{\mathfrak{I}} \text{ and } (e, e^*) \in l^{\mathfrak{I}}\}$
- $(\exists U_1, U_2.r)^{\mathfrak{I}} = \{d \in \Delta^{\mathfrak{I}} \mid \text{there exist } e_1^*, e_2^* \text{ s.t. } (d, e_1^*) \in U_1^{\mathfrak{I}}, (d, e_2^*) \in U_2^{\mathfrak{I}} \text{ and } (e_1^*, e_2^*) \in r^{\mathfrak{I}}\}$
- $(\forall U_1, U_2.r)^{\mathfrak{I}} = \{d \in \Delta^{\mathfrak{I}} \mid \text{for all } e_1^*, e_2^* \text{ s.t. } (d, e_1^*) \in U_1^{\mathfrak{I}}, (d, e_2^*) \in U_2^{\mathfrak{I}} \text{ it holds that } (e_1^*, e_2^*) \in r^{\mathfrak{I}}\}$

Now I can define the following combined spatio-thematic logic, where a^*, b^* stand for constants intended to be interpreted by regions:

Definition 4.1 (DL-Lite$^{\sqcap}_{\mathcal{F},\mathcal{R}}$(RCC8)) Let $r \in Rel_{RCC8}$ and $T_\omega = Ax_{RCC8}$.

roles(σ):	$R \longrightarrow P \mid P^-$	
paths(σ)	$U \longrightarrow R \mid \tilde{R}$	
concepts(σ):	$B \longrightarrow A \mid \exists R \mid \exists l$	(basic concepts)
	$C_l \longrightarrow B \mid C_l \sqcap B$	(concepts on lhs)
	$C_r \longrightarrow B \mid \neg B \mid \exists U_1, U_2.r$	(concepts on rhs)
tbox-axioms(σ):	$C_l \sqsubseteq C_r$, (funct l), (funct R), $R_1 \sqsubseteq R_2$	
abox-axioms(σ):	$A(a), R(a, b), l(a, a^*), r(a^*, b^*)$	
Constraint:	If (funct R) $\in \mathcal{T}$, then R and R^- do not occur on rhs of a role inclusion axiom or in a concept of the form $\exists U_1, U_2.r$	

As satisfiability checking of RCC8 constraint networks is NPTIME-complete, there is only a chance to reach FOL rewritability if the constraint network in the abox is consistent and complete, i.e., if the constraint network has exactly one solution and if, in addition, it is a clique with base relations as labels. In this case the abox is called *spatially complete*. For cadastral maps or maps containing areas of administration one can assume pretty safely (almost) spatial completeness. The coupling with RCC8 is so weak that FOL rewritability of satisfiability follows.

Proposition 4.2 *Checking satisfiability of DL-Lite*$^{\sqcap}_{\mathcal{F},\mathcal{R}}$*(RCC8) ontologies that have a spatially complete abox is FOL rewritable.*

Proof See p. 91. □

Testing whether FOL rewritability holds for satisfiability tests is necessary for tests whether FOL rewritability is provable for query answering w.r.t. a sufficiently expressive query language. The query language which I consider is derived from grounded conjunctive queries and is denoted by GCQ^+. This query language is explicitly constructed for use with DL-Lite$^{\sqcap}_{\mathcal{F},\mathcal{R}}(RCCi)$ and DL-Lite$^{\sqcap,+}_{\mathcal{F},\mathcal{R}}(RCCi)$ (for $i \in \{2, 3, 5, 8\}$) and so provides only means for qualitative spatial queries. But it could be extended to allow also for quantitative spatial queries.

Definition 4.3 Let $\mathcal{L}_i \in \{$DL-Lite$^{\sqcap}_{\mathcal{F},\mathcal{R}}(RCCi)$, DL-Lite$^{\sqcap,+}_{\mathcal{F},\mathcal{R}}(RCCi)$ $| \in \{2, 3, 5, 8\}\}$. A GCQ^+ *atom w.r.t.* $\mathcal{L}_i$ is a formula of one of the following forms:

- $C(x)$, where C is a $\mathcal{L}_i$ concept without the negation symbol and x is a variable or a constant.
- $(\exists R_1 \dots R_n.C)(x)$ for role symbols or their inverses R_i, a DL-Lite$^{\sqcap}_{\mathcal{F},\mathcal{R}}$ (RCCi) concept C without the negation symbol, and a variable or a constant x
- $R(x, y)$ for a role symbol R or an inverse thereof
- $l(x, y^*)$, where x is a variable or constant and y^* is a variable or constant intended to denote elements of models Ax_{RCCi}
- $r(x^*, y^*)$, where $r \in Rel_{RCCi}$ and x^*, y^* are variables or constants intended to denote elements of models Ax_{RCCi}

A GCQ^+ *query w.r.t.* $\mathcal{L}_i$ is a query $\tilde{\exists}\mathbf{y}\mathbf{z}^* \bigwedge C_i(\mathbf{x}, \mathbf{w}^*, \mathbf{y}, \mathbf{z}^*)$ where all $C_i(\mathbf{x}, \mathbf{w}^*, \mathbf{y}, \mathbf{z}^*)$ are GCQ^+ atoms and $\tilde{\exists}\mathbf{y}\mathbf{z}^* = \tilde{\exists}y_1 \dots \tilde{\exists}y_n\tilde{\exists}z_1^* \dots \tilde{\exists}z_m^*$ is a sequence of $\exists$-quantifiers interpreted w.r.t. the active domain semantics.

The perfect-rewriting algorithm presented in the following is an adaptation of the algorithm PerfectRef [4, Fig. 13] for reformulating UCQs w.r.t. DL-Lite ontologies to the spatio-thematic setting in which GCQ^+-queries are asked over $\mathcal{L}_i$ ontologies.

Given a query GCQ^+ $\phi(\mathbf{x})$, I transform it to a special form. $\tau_1(\phi(\mathbf{x}))$ is the result of the transformation to a UCQ and $\tau_2(\phi(\mathbf{x}))$ is the result of transforming $\phi(\mathbf{x})$ in a hybrid UCQ whose conjuncts are either classical predicate logical atoms or GCQ^+-atoms which are not further transformed. I use the notation "$g = F$" for "g is of the form F".

The original algorithm PerfectRef operates on the positive inclusion (PI) axioms of a DL-Lite ontology by using them as rewriting aids for the atomic formulas in the

input : a hybrid query $\tau_1(\phi(\mathbf{x})) \cup \tau_2(\phi(\mathbf{x}))$, DL-Lite(RCC8) tbox $\mathcal{T}$
output : a UCQ pr

1 $pr := \tau_1(\phi(\mathbf{x})) \cup \tau_2(\phi(\mathbf{x}))$;
2 **repeat**
3     $pr' := pr$;
4   **foreach** *query* $q' \in pr'$ **do**
5   **foreach** *atom* g *in* q' **do**
6   **if** g *is a FOL-atom* **then**
7   **foreach** *PI* α *in* $\mathcal{T}$ **do**
8   **if** α *is applicable to* g **then**
9                         $pr := pr \cup \{q'[g/gr(g, \alpha)]\}$;
10   **end**
11   **end**
12   **else**
13   **if** $g = \exists\tilde{R}_1, \tilde{R}_2.r_3(x)$ *and* $r_1; r_2 \subseteq r_3$ **then**
14                     $X := q'[g/(\exists\tilde{R}_1, l.r_1(x) \wedge \exists l, \tilde{R}_2.r_2(x))]$;
15                     $pr := pr \cup \{X\} \cup \{\tau_2(X, \{\exists\tilde{R}_1, l.r_1(x), \exists l, \tilde{R}_2.r_2(x)\})\}$
16   **end**
17   **if** $g = \exists U_1, U_2.r_1(x)$ *and* $B \sqsubseteq \exists U_1, U_2.r_2(x) \in \mathcal{T}$ *for* $r_2 \subseteq r_1$ **then**
18                     $pr := pr \cup \{q'[g/B(x)]\}$;
19   **end**
20   **if** $g = \exists U_1, U_2.r_1(x)$ *and* $B \sqsubseteq \exists U_1, U_2.r_2(x) \in \mathcal{T}$ *for* $r_2^{-1} \subseteq r_1$ **then**
21                     $pr := pr \cup \{q'[g/B(x)]\}$;
22   **end**
23   **if** $g = \exists\tilde{R}_1, U_1.r(x)$ *(resp.* $\exists U_1, \tilde{R}_1.r(x)$*) and (*$R_2 \sqsubseteq R_1 \in \mathcal{T}$ *or* $R_2^{-1} \sqsubseteq R_1^{-1} \in \mathcal{T}$*)* **then**
24                     $X := q'[g/(g[R_1/R_2])]$;
25                     $pr := pr \cup \{X\} \cup \{\tau_2(X, \{g[R_1/R_2]\})\}$;
26   **end**
27   **end**
28   **end**
29   **foreach** *pair of FOL-atoms* g_1, g_2 *in* q' **do**
30   **if** g_1 *and* g_2 *unify* **then**
31                 $pr := pr \cup \{anon(reduce(q', g_1, g_2))\}$;
32   **end**
33   **end**
34   **end**
35 **until** $pr' = pr$;
36 **return** $drop(pr)$

Algorithm 1: Adapted PerfectRef

UCQ. Lines 5–12 and 28–34 of my adapted algorithm (Algorithm 1) make up the original PerfectRef. Roughly, the PerfectRef algorithm acts in the inverse direction with respect to the chasing process. For example, if the tbox contains the PI axiom $A_1 \sqcap A_2 \sqsubseteq A_3$, and the UCQ contains the atom $A_3(x)$ in a CQ, then the new rewritten UCQ query contains a CQ in which $A_3(x)$ is substituted by $A_1(x) \wedge A_2(x)$.

The applicability of a PI axiom to an atom is restricted in those cases where the variables of an atom are either distinguished variables or also appear in another atom of the CQ at hand. To handle these cases, PerfectRef—as well as also my adapted version—uses anonymous variables _ to denote all non-distinguished variables in an atom that do not occur in other atoms of the same CQ. The function anon (line 31 in Algorithm 1) implements the anonymisation. The application conditions for PI axioms α and atoms are as follows: α is applicable to $A(x)$ if A occurs on the right-hand side; and α is applicable to $R(x_1, x_2)$, if $x_2 = \_$ and the right-hand side of α is $\exists R$; or $x_1 = \_$ and the right-hand side of α is $\exists R^-$; or α is a role inclusion assertion and its right-hand side is either R or R^-. The outcome $gr(g, \alpha)$ of applying an applicable PI α to an atom g corresponds to the outcome of resolving α with g. For example, if α is $A \sqsubseteq \exists R$ and g is $R(x, \_)$, the result of the application is $gr(g, \alpha) = A(x)$. I leave out the details [4, Fig.12, p. 307]. In PerfectRef, atoms in a CQ are rewritten with the PI axioms (lines 6–11) and, if possible, merged by the function reduce (line 31) which unifies the atoms with the most general unifier (lines 28–34).

The modification of PerfectRef concerns the handling of GCQ^+-atoms of the form $\exists U_1, U_2.r(x)$. These atoms may have additional implications that are accounted for with four cases (lines 12–26 of the algorithm). At the end of the adapted algorithm PerfectRef (Algorithm 1, line 35) these atoms are deleted by calling the function *drop*. The algorithm returns a classical UCQ, which can be evaluated as an SQL query on the database $DB(\mathcal{A})$.

That the rewriting given in Algorithm 1 is indeed correct and complete follows from Theorem 4.4.

Theorem 4.4 *Answering GCQ^+-queries w.r.t. DL-Lite$^{\sqcap}_{\mathcal{F},\mathcal{R}}$(RCC8) ontologies that have a spatially complete abox is FOL rewritable.*

Proof See p. 92 □

4.3 Example Scenario

In order to illustrate the spatio-thematic lightweight logics of this chapter, I describe a simple application scenario in which an engineering bureau plans additional parks in a city [16]. Assume, the bureau has stored geographical data in some database (DB) and declares relevant concepts in the terminological part of his knowledge base, the tbox. The engineer gives necessary conditions for a concept *Park+Lake* which is a park containing a lake that touches it from within, i.e., using the terminology of the region connection calculus (RCC) [19], the lake is a tangential proper part of the park. Similarly, a necessary condition for the concept *Park4Playing* is given which is a park containing a playing ground (for children) that is a tangential proper part.

I assume that the data are mapped to an abox, the logical pendant of the DB. In particular the data should generate the fact that there is an object a which is both a park with a lake and with a playing area, that is *Park+Lake*(a) and *Park4Playing*(a) are contained in the abox. But the location of a is not known. Think of a as an object whose architectural design is determined but the place where a is going to be localised is not determined yet.

Now, the engineering bureau asks for all parks with lakes and playing areas such that the playing area is not contained as island in the lake. These kinds of parks can be thought of as safe as the playing ground can be directly reached from the park (without a bridge). All objects that fall into the answer set of this query w.r.t. the tbox and the data can have one of the configurations A to C illustrated in Fig. 4.1 (and many more) but are not allowed to have the configuration D. The object a has to be in the answer set to the original query as the tbox together with the abox and some deduction on the spatial configuration entails that a is an object which excludes the island configuration D. Remember that a is "abstract" in so far as its geographical location is not known. So in fact deduction is needed to see that a does not have configuration D.

The knowledge in this scenario is captured in the following DL-Lite$^{\sqcap}_{\mathcal{F},\mathcal{R}}$ (RCC8) ontology: The tbox $\mathcal{T}$ of the engineering bureau contains the following DL-Lite$^{\sqcap}_{\mathcal{F},\mathcal{R}}$ (RCC8) axioms:

$$\begin{aligned} \mathit{Park{+}Lake} &\sqsubseteq \mathit{Park} \\ \mathit{Park4Playing} &\sqsubseteq \mathit{Park} \\ \mathit{Park{+}Lake} &\sqsubseteq \exists \mathit{hasLake} \circ l, l.\mathsf{tpp} \\ \mathit{Park4Playing} &\sqsubseteq \exists \mathit{hasPlAr} \circ l, l.\mathsf{tpp} \end{aligned}$$

The abox $\mathcal{A}$ contains at least the following axioms

$$\mathit{Park{+}Lake}(a), \mathit{Park4Playing}(a)$$

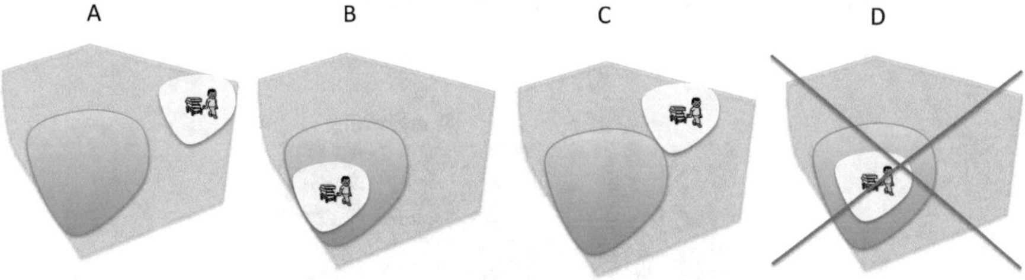

Fig. 4.1 The first three pictures show allowed spatial configurations in the engineering bureau scenario. The last depicts a forbidden one.

The query of the engineer, which asks for all parks with lakes and playing areas such that the playing area is not a tangential proper part of the lake, can be formalised by the following GCQ^+ query:

$$\alpha_0(x) = \mathit{Park}(x) \land \exists \mathit{hasLake} \circ l, \mathit{hasPlAr} \circ l.(\mathcal{B}_{RCC8} \setminus \{\mathsf{ntpp}\})(x)$$

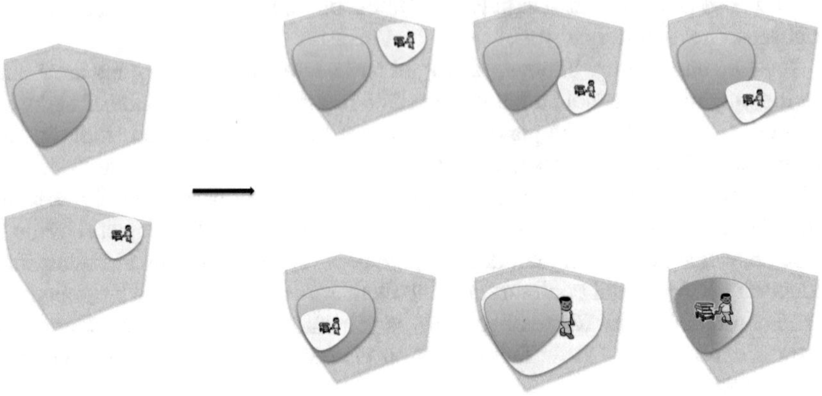

Fig. 4.2 Illustration for composition entry tpp; tppi.

The main step requiring spatial reasoning is that of incorporating the composition entry for tpp; tppi which is formally given by

$$\text{tpp; tppi} = \{\text{dc, ec, po, tpp, tppi, eq}\} \subseteq \mathcal{B}_{RCC8} \setminus \{\text{ntpp}\}$$

Figure 4.2 illustrates the composition entry w.r.t. this scenario. Here x corresponds to the lake, y to a park and z to a playing ground for children. Figure 4.3 describes the interaction of the RCC component with the thematic component.

Using the composition entry for tpp; tppi, the reformulation algorithm introduced above (lines 13–15) produces a UCQ that contains the following CQ:

$$\alpha_1(x) = (\exists hasLake \circ l, l.\text{tpp})(x) \wedge (\exists l, hasPlAr \circ l.\text{tppi})(x)$$

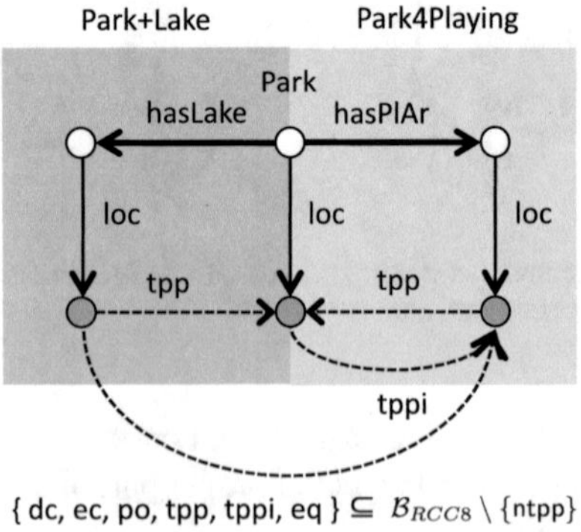

Fig. 4.3 Interpretation satisfying the original query and its rewritings

Rewriting $\exists l, hasPlAr \circ l.\mathsf{tppi}$ to $\exists hasPlAr \circ l, l.\mathsf{tpp}$ (lines 20–21) in combination with the rewriting rule for $A_1 \sqsubseteq A_2$ (Def. 2.10) results in another CQ:

$$\alpha_2(x) = Park{+}Lake(x) \wedge Park4Playing(x)$$

The resulting rewritten query $\alpha_{rew}(x)$ is

$$\alpha_{rew}(x) = \alpha_0(x) \vee \alpha_1(x) \vee \alpha_2(x)$$

It captures (as desired) the object a.

4.4 Strong Combinations of DL-Lite with RCC

Another way of reaching FOL rewritability for combinations of DL-Lite with RCC is weakening the expressivity of the spatial component. Hence, one may ask whether a combination of DL-Lite with the calculus RCC3 or RCC2 [24], both fragments with weak expressibility, allows for FOL rewritability w.r.t. satisfiability checks (and query answering). Their potential use as logics for approximating [9] ontologies in more expressible combined logics such as $\mathcal{ALC}$(RCC8) makes the investigation valuable (see also the résumé of this chapter). The logics DL-Lite$^{\sqcap,+}_{\mathcal{F},\mathcal{R}}$(RCC2) and DL-Lite$^{\sqcap,+}_{\mathcal{F},\mathcal{R}}$(RCC3) are defined as follows ('+' indicates the strong combination):

Definition 4.5 (DL-Lite$^{\sqcap,+}_{\mathcal{F},\mathcal{R}}$(RCC2) and DL-Lite$^{\sqcap,+}_{\mathcal{F},\mathcal{R}}$(RCC3)) Let $T_\omega = Ax_{RCC2}$ resp. $T_\omega = Ax_{RCC3}$ and $r \in \mathcal{B}_{RCC2}$ resp. $r \in \mathcal{B}_{RCC3}$

roles(σ):	$R \longrightarrow P \mid P^-$	
paths(σ)	$U \longrightarrow l \mid \tilde{R}$	
concepts(σ):	$B \longrightarrow A \mid \exists R$	(basic concepts)
	$C_l \longrightarrow B \mid C_l \sqcap B$	(concepts on lhs)
	$C_r \longrightarrow B \mid \neg B \mid \exists U_1, U_2.r$	(concepts on rhs)
tbox-axioms(σ):	$C_l \sqsubseteq C_r$, (funct l, R), $R_1 \sqsubseteq R_2$	
abox-axioms(σ):	$A(a), R(a,b), l(a,a^*), r(a^*,b^*)$	
Constraint:	If (funct R) $\in \mathcal{T}$, then R and R^- do not occur on the rhs of a role inclusion axiom	

For RCC3, the strong combination with DL-Lite$^{\sqcap}_{\mathcal{F},\mathcal{R}}$ leads to non-FOL rewritability. The reason lies in the fact that testing the satisfiability of RCC3 is not in the complexity class AC^0. This in turn is a consequence of the following lemma which uses the fact that the reachability problem in symmetric (undirected) graphs is logspace complete [20].

Lemma 4.6 *Checking satisfiability of RCC3 networks is LOGSPACE hard.*

Proof See p. 95 □

This lemma immediately entails the fact that satisfiability checking for ontologies over the logic DL-Lite$^{\sqcap,+}_{\mathcal{F},\mathcal{R}}$(RCC3) is not FOL rewritable. This problem does not vanish if abox $\mathcal{A}$ is assumed to be spatially complete—as shown by the following proposition.

Proposition 4.7 *Satisfiability checking of ontologies in DL-Lite$^{\sqcap,+}_{\mathcal{F},\mathcal{R}}$(RCC3) with spatially complete aboxes is not FOL rewritable.*

Proof See p. 95

As a corollary to this proposition it follows that strong combinations of DL-Lite with RCC5 and RCC8—denoted DL-Lite$^{\sqcap,+}_{\mathcal{F},\mathcal{R}}$(RCC5) and DL-Lite$^{\sqcap,+}_{\mathcal{F},\mathcal{R}}$(RCC8), respectively defined in the same manner as in Definition 4.5—do not allow for FOL rewriting of satisfiability checking.

Corollary 4.8 *DL-Lite$^{\sqcap,+}_{\mathcal{F},\mathcal{R}}$(RCC5) and DL-Lite$^{\sqcap,+}_{\mathcal{F},\mathcal{R}}$(RCC8) do not allow for FOL rewriting of satisfiability checking.*

The low resolution calculus RCC2 is quite more inexpressive than RCC3 due to the fact that the composition table does not allow for the propagation of information: all compositions of DR, O result in the maximally unspecified relation $\{\mathsf{DR}, \mathsf{O}\}$. Hence, FOL rewritability of satisfiability testing follows easily considering the query $\phi() = \exists x, y[\mathsf{O}(x, y) \wedge \mathsf{DR}(x, y)] \vee \exists x[\mathsf{DR}(x, x)]$.

Proposition 4.9 *Testing the satisfiability of RCC2 networks is FOL rewritable.*

But in combination with functionality axioms of the tbox one could have the problem that the abox may lead to identifications of regions. The identified regions are not allowed to have edges labelled O, DR, resp. Though this can be tested, the problem arises when a chain of regions is identified by the tbox and the abox, because the length of the chain is not known in advance. This is explained more formally as follows: in addition to RCC2 constraint-network assertions I allow identity assertions $v = w$ for regions v, w. As one can assume that all nodes in a RCC2 network are connected by an edge labelled O, DR or $\mathcal{B}_{RCC2}$, I use a more intuitive formalism where, for every assertion $v = w$, the label of the edge between v and w is marked with an =. For example, an edge between v, w with label $\mathsf{DR}^=$ stands for $\mathsf{DR}(v, w) \wedge v = w$. I call such a network an =-marked RCC2 network (a RCC$^=$2 network for short). Let $\mathcal{B} = \mathcal{B}_{RCC2}$ in the following.

Proposition 4.10 *An RCC$^=$2 constraint network N is unsatisfiable iff*

1. *N contains $\mathsf{DR}(v, v)$ or $\mathsf{DR}^=(v, v)$ for some node v; or*
2. *N contains $\mathsf{DR}^=(v, w)$; or*
3. *N contains a cycle in which there is $\mathsf{DR}(v, w)$ and in which there is a path from v to w such that every label on the path is $\mathcal{B}^=$ or $\mathsf{O}^=$; or*
4. *N contains a cycle in which there is $\mathsf{DR}(v, w)$ and in which there is a path from v to w s.t. every label on the path is $\mathcal{B}^=$ or $\mathsf{O}^=$ except one which is O.*

Proof See p. 96. □

Proposition 4.10 shows that adding identity assertions to an RCC2 network may require checking the existence of identity chains of arbitrary length. Hence, in principle it is possible that the functional roles used in DL-Lite$^{\sqcap,+}_{\mathcal{F},\mathcal{R}}$(RCC2) may lead to identity chains. But as the following proposition shows, this cannot be the case: the identity paths induced by functionalities in DL-Lite$^{\sqcap,+}_{\mathcal{F},\mathcal{R}}$(RCC2) can have only a maximal length of one.

Proposition 4.11 *Satisfiability checking of ontologies in DL-Lite$^{\sqcap,+}_{\mathcal{F},\mathcal{R}}$(RCC2) is FOL rewritable.*

Proof See p. 97. □

4.5 Related Work

As the logics discussed here are derived from the framework of [11], it has to be mentioned as the most relevant for the work described in this chapter. The main difference is that, here, I considered fragments of the logic of [11] under FOL rewritability aspects and, further, that I worked with ω-admissible theories and not ω-admissible concrete domains.

Other work combining descriptions logics and spatial calculi can be found in [7, 24]. But none of the logics described there aim at FOL rewritability.

Whereas the idea in the papers mentioned above is to build combinations of logics for spatio-thematic reasoning, early investigations on RCC aimed on finding appropriate logics that semantically found and enable RCC reasoning. Here one should mention the translations of RCC into modal logic as given by [2] and later by [14].

My own contribution [18] (which is not discussed in this monograph) combines DLs and spatial concepts in the sense that it considers convex regions as potential extensions for concepts—thereby providing a spatial semantics for DLs.

Also related to the work presented in this chapter is the work on constraint-based spatial reasoning [21]. It offers a well-developed and well proven theory for spatial domains, but it does not fill in the need for a system that combines reasoning over a spatial and a non-spatial (thematic) domain. Though constraint databases [10] are good candidate frameworks for reasoning over a mixed domain of geo-thematic objects, the investigations on constraint databases so far did not incorporate terminological reasoning in the paradigm of OBDA. Maybe, when constraint databases become en vogue again, work on OBDA for constraint DBs will become worth pursuing.

4.6 Summary

Combining DL-Lite with expressive fragments of the region calculus such as RCC8 into logics that preserve the property of FOL rewritability is possible if the coupling is weak, i.e., if constraints of the RCC8 network contained in the abox are not transported over to the implicitly constructed constraint network resulting from the constructors of the form $\exists U_1, U_2.r$. On the other hand, one may try to aim for stronger combinations by using weaker calculi such as RCC2 or RCC3. As was shown by a reduction proof, a strong combination with RCC3 destroys the FOL rewritability of satisfiability checking. The reason is that checking the satisfiability of RCC3 networks needs to test for reachability along EQ paths, which can be reproduced by the tbox. For the low resolution calculus RCC2, FOL rewritability of satisfiability checking is provable—though checking the satisfiability of RCC2 networks with additional identity assertions is at least as hard as checking RCC3 networks.

The resulting combined spatio-thematic logics allowing for FOL rewriting, i.e., DL-Lite$^{\sqcap,+}_{\mathcal{F},\mathcal{R}}$(RCC2) and DL-Lite$^{\sqcap}_{\mathcal{F},\mathcal{R}}$(RCC8) are very weak in expressivity, and so one might ask whether they can be used at all for interesting (GIS) applications. In principle they can be considered as potential candidates for approximation. If one represents the spatio-thematic knowledge in an expressive logic $\mathcal{L}_{expr}(RCC8)$ such as $\mathcal{ALC}(RCC8)$ and wants to apply query answering over geographical data, then one possible way is to approximate the ontology with an ontology in a weaker spatio-thematic logic. More concretely, assume that $\mathcal{T} \cup \mathcal{A}$ is some ontology in a highly expressive spatio-thematic logic $\mathcal{L}_{expr}$(RCC8), say $\mathcal{ALC}(RCC8)$. One considers a tbox $\mathcal{T}'$ in a weak spatio-thematic logic (such as DL-Lite$^{\sqcap,+}_{\mathcal{F},\mathcal{R}}$(RCC2)) that approximates $\mathcal{T}$. There are two different approximation methods: the approximation may be from below (as in [5]), which means that $\mathcal{T} \models \mathcal{T}'$, or it may be from above as in [9], which means that $\mathcal{T}' \models \mathcal{T}$.

The consequence for the approximation from below is that the set of answers to a query ϕ with respect to the approximating ontology $\mathcal{T}' \cup \mathcal{A}$ is a subset of the answers with respect to the approximated ontology $\mathcal{T} \cup \mathcal{A}$, i.e., the set of answers is sound but not complete. There is no further way to complete the answer set (other than running a full reasoner over the original ontology) and hence the only useful action after approximation is to inform the user about the tbox axioms that have been changed.

In case of approximation from above the consequences are the following: The approximation entails that all answers to the original query ϕ with respect to $\mathcal{T} \cup \mathcal{A}$ is a subset of the answers with respect to $\mathcal{T}' \cup \mathcal{A}$. Hence the approximation leads to complete but not necessarily correct query answering. This in turn means that the set of answers $cert(\phi, \mathcal{T}' \cup \mathcal{A})$ has to be verified in a post-processing phase with a full reasoner that can process ontologies in $\mathcal{L}_{expr}$(RCC8). Note that there is no way around using a reasoner also in case of approximating from above but the main difference is that one has a possibly small set of instance queries with concretely named instances instead of the original retrieval query. In many cases, this enables the use of optimisations (such as those based on abox modularisation [13]).

Regarding the negative results proved in this chapter, a natural question is whether there are other ways to achieve spatio-thematic logics that can be used for strict OBDA. The following list gives a tentative answer with pointers to further research questions.

- First of all one might change the type of spatial calculus. In this chapter this was done only locally, looking at different RCC calculi RCCi. But all of them were RCC calculi. The main problem with these is the non-guarded use of a binary connectedness relation—which does not fit to the concept-oriented and hence tree-structured modelling methodology of description logics. An interesting alternative is the theory of spatial relation based on partition chains as developed in Chap. 3. Due to the underlying tree structured relation the strong combinations of DL-Lite seem to be promising.
- The way of combining the thematic and the spatial component might be different from the way of combination proposed by [11] which was also used in this chapter. Instead of new quantifiers with property paths, an alternative, more homogeneous integration could be sought for: instead of combining two logics one searches for an integrated spatio-thematic logic.
- Another possibility is to use additional knowledge of the properties of the abox that one has in advance, i.e., before the rewriting. Some form of knowledge along this line has already been used in this chapter when I assumed that the abox is spatially complete. But of course further conditions could be considered that are (usually) fulfilled by real-world databases and the induced aboxes. This is similar to the approach of parameterised complexity [6] where the idea is to identify parameters that are responsible for the non-feasibility of given algorithmic problems. Knowing a bound on these parameters makes the problems feasible. A typical example in graph algorithms is the knowledge about the bounded tree-width of graphs. For example, as shown in [3], deciding the satisfiability of an RCC8 constraint network with bounded tree-width can be done in polynomial time.
  Also the idea of combined rewriting as outlined by [12] can be considered as incorporating knowledge of the abox in advance. Here one can use the abox to merge it with the tbox into some interpretation that can be used as a universal model. This model is then used as the database on which the rewritten queries answered. So in case of the combined rewriting method, one has total knowledge on the abox before the rewriting step. In this case, the restriction to DL-Lite is not necessary anymore so that one can consider other light-weight DLs such as those from the $\mathcal{EL}$ family [1].
  The use of integrity constraints for mappings developed in the thesis of [22] is another way to convey information from the abox.
- Another possibility is to constrain the source query language, i.e., the query using the signature of the spatio-thematic ontology. Sometimes it suffices to consider only tree-shaped or acyclic conjunctive queries.
- A last aspect concerns the target query language into which the source queries are rewritten. Choosing a more expressive query language (such as datalog, or FOL extended with counting quantifiers etc.) may allow for rewriting where FOL

does not. This possibility depends of course on the fact whether the backend repository supports the language at all and whether it provides a feasible means of answering queries formulated in that language.

References

1. Baader, F., Brandt, S., Lutz, C.: Pushing the EL envelope. In: Proceedings of the 19th International Joint Conference on Artificial Intelligence (IJCAI-05), pp. 364–369. Morgan Kaufmann Publishers Inc., San Francisco, CA, USA (2005)
2. Bennett, B.: Modal logics for qualitative spatial reasoning. Logic Journal of the IGPL **4**(1), 23–45 (1996)
3. Bodirsky, M., Wölfl, S.: RCC8 is polynomial on networks of bounded treewidth. In: Proceedings of the 22nd International Joint Conference on Artificial Intelligence (IJCAI-11), *IJCAI'11*, vol. 2, pp. 756–761. AAAI Press (2011). DOI 10.5591/978-1-57735-516-8/IJCAI11-133
4. Calvanese, D., De Giacomo, G., Lembo, D., Lenzerini, M., Poggi, A., Rodríguez-Muro, M., Rosati, R.: Ontologies and databases: The DL-Lite approach. In: Proceedings of the 5th International Reasoning Web Summer School (RW-09), *LNCS*, vol. 5689, pp. 255–356. Springer (2009)
5. Console, M., Santarelli, V., Savo, D.F.: From owl to DL-Lite through efficient ontology approximation. In: W. Faber, D. Lembo (eds.) Proceedings of the 7th International Conference on Web Reasoning and Rule Systems (RR-13), pp. 229–234. Springer Berlin Heidelberg, Berlin, Heidelberg (2013). DOI 10.1007/978-3-642-39666-3_20
6. Flum, J., Grohe, M.: Parameterized Complexity Theory. Texts in Theoretical Computer Science. An EATCS Series. Springer Berlin Heidelberg (2006)
7. Haarslev, V., Lutz, C., Möller, R.: A description logic with concrete domains and a role-forming predicate operator. Journal of Logic and Computation **9**(3), 351–384 (1999)
8. Johnson, A.G., Galdi, D.E., Godwin, L., Franz, L., Boudriault, G., Ratcliffe, M.R.: Tiger/Line® shapefiles. Technical Documentation prepared by the US Census Bureau (2009)
9. Kaplunova, A., Moeller, R., Wandelt, S., Wessel, M.: Towards scalable instance retrieval over ontologies. In: B. Yaxin, W. Mary-Anne (eds.) Proceedings of the 4th International Conference on Knowledge Science, Engineering and Management (KSEM-10), *LNCS*, vol. 6291. Springer (2010)
10. Kuper, G.M., Libkin, L., Paredaens, J. (eds.): Constraint Databases. Springer (2000)
11. Lutz, C., Milicic, M.: A tableau algorithm for description logics with concrete domains and general TBoxes. Journal of Automated Reasoning **38**(1-3), 227–259 (2007)
12. Lutz, C., Toman, D., Wolter, F.: Conjunctive query answering in the description logic $\mathcal{EL}$ using a relational database system. In: Proceedings of the 21st International Joint Conference on Artificial Intelligence (IJCAI-09). AAAI Press (2009)
13. Möller, R., Neuenstadt, C., Özçep, Ö.L., Wandelt, S.: Advances in accessing big data with expressive ontologies. In: Proceedings of the 36th Annual German Conference on Artificial Intelligence (KI-13), *LNCS*, vol. 8077, pp. 118–129. Springer (2013)
14. Nutt, W.: On the translation of qualitative spatial reasoning problems into modal logics. In: W. Burgard, A. Cremers, T. Cristaller (eds.) Proceedings of the 23rd Annual German Conference on Artificial Intelligence (KI-99), *LNCS*, vol. 1701, pp. 699–699. Springer Berlin / Heidelberg (1999)
15. Özçep, Ö.L., Möller, R.: Combining DL-Lite with spatial calculi for feasible geo-thematic query answering. In: Y. Kazakov, D. Lembo, F. Wolter (eds.) Proceedings of the 25th International Workshop on Description Logics (DL-12), vol. 846 (2012)
16. Özçep, Ö.L., Möller, R.: Computationally feasible query answering over spatio-thematic ontologies. In: Proceedings of the 4th International Conference on Advanced Geographic Information Systems, Applications, and Services (GEOProcessing-12) (2012)

17. Özçep, Ö.L., Möller, R.: Scalable geo-thematic query answering. In: P. Cudré-Mauroux, J. Heflin, E. Sirin, T. Tudorache, J. Euzenat, M. Hauswirth, J.X. Parreira, J. Hendler, G. Schreiber, A. Bernstein, E. Blomqvist (eds.) Proceedings of the 11th International Semantic Web Conference (ISWC-12), vol. 7649, pp. 658–673 (2012)
18. Özçep, Ö.L., Möller, R.: Spatial semantics for concepts. In: T. Eiter, B. Glimm, Y. Kazakov, M. Krötzsch (eds.) Proceedings of the 26th International Workshop on Description Logics (DL-13), *CEUR Workshop Proceedings*, vol. 1014, pp. 816–828. CEUR-WS.org (2013)
19. Randell, D.A., Cui, Z., Cohn, A.G.: A spatial logic based on regions and connection. In: Proceedings of the 3rd International Conference on Knowledge Representation and Reasoning (KR-92), pp. 165–176 (1992)
20. Reingold, O.: Undirected connectivity in log-space. Journal of the ACM **55**(4), 17:1–17:24 (2008). DOI 10.1145/1391289.1391291
21. Renz, J., Nebel, B.: Qualitative spatial reasoning using constraint calculi. In: M. Aiello, I. Pratt-Hartmann, J. Benthem (eds.) Handbook of Spatial Logics, pp. 161–215. Springer Netherlands (2007)
22. Rodríguez-Muro, M.: Tools and techniques for ontology based data access in lightweight description logics. Ph.D. thesis, Free University of Bozen - Bolzano (2010)
23. Rodriguez-Muro, M., Calvanese, D.: Semantic index: Scalable query answering without forward chaining or exponential rewritings. In: Posters of the 10th International Semantic Web Conference (ISWC-11) (2011)
24. Wessel, M.: Qualitative spatial reasoning with the $\mathcal{ALCI}_{RCC}$- family — first results and unanswered questions. Technical Report FBI–HH–M–324/03, University of Hamburg, Department for Informatics (2003)

Appendix

Proof of Proposition 4.2

Let $\mathcal{T} \cup \mathcal{A} \cup Ax_{RCC8}$ be an ontology with a spatially complete abox $\mathcal{A}$ and the ω-admissible background theory Ax_{RCC8}. I build a simple closure $\mathcal{T}'$ of the pure DL-Lite part of $\mathcal{T}$ in the following way. Every DL-Lite axiom of $\mathcal{T}$ is in $\mathcal{T}'$. For every $B \sqsubseteq \exists \tilde{R}_1, \tilde{R}_2.r \in \mathcal{T}$ let $\{B \sqsubseteq \exists R_1, B \sqsubseteq \exists R_2\} \subseteq \mathcal{T}'$, for $B \sqsubseteq \exists \tilde{R}, l.r \in \mathcal{T}$ and $B \sqsubseteq \exists l, \tilde{R}.r \in \mathcal{T}$ let $\{B \sqsubseteq \exists R, B \sqsubseteq \exists l\} \subseteq \mathcal{T}'$. I claim that $\mathcal{T} \cup \mathcal{A} \cup Ax_{RCC8}$ is satisfiable iff the DL-Lite ontology $\mathcal{T}' \cup (\mathcal{A} \setminus N_{\mathcal{A}})$ is satisfiable. Here $N_{\mathcal{A}}$ denotes the RCC8-network contained in $\mathcal{A}$. The difficult direction is the one from right to left which I will prove in the following. Let $\mathcal{T}' \cup (\mathcal{A} \setminus N_{\mathcal{A}})$ be satisfiable by some model $\mathfrak{I}$ which can w.l.o.g. be assumed to be the canonical model for the chase $chase(\mathcal{T}_p' \cup (\mathcal{A} \setminus N_{\mathcal{A}}))$. Here, $\mathcal{T}_p'$ denotes the PI axioms in $\mathcal{T}$. As $\mathcal{A}$ is spatially complete, there exists a model $\mathfrak{I}' \models N_{\mathcal{A}} \cup Ax_{RCC8}$. I let $X = chase(\mathcal{T}_p' \cup \mathcal{A})$ which is satisfiable by an interpretation $\mathfrak{J}$ built as a merge of $\mathfrak{I}$ and $\mathfrak{I}'$. Now, I will extend X by further chasing steps in the following way. Different from the chase construction of [4], I start with the set X which may already be infinite. This fact poses no problems as I chose an ordering over set of all possible strings over the signature of the ontology and the chasing constants.

In addition to the chasing rules listed in Definition 2.10, I will use a chasing rule for axioms of the form $B \sqsubseteq \exists \tilde{R}_1, l.(r_1 \vee \cdots \vee r_k) \in \mathcal{T}$ and the other axioms of the form $B \sqsubseteq \exists U_1, U_2.R \in \mathcal{T}$ (Fig. 4.4). Let S_i denote the sets created during the chasing process. Directly after applying this chasing rule a completion step is applied in order to make the generated constraint network have a unique model modulo isomorphism. For every node n^* appearing in S_i an atom $r_{n*}(n^*, y^*)$, where $r_{n^*} \in \mathcal{B}_{RCC8}$ is some basic relation, is added. There may be infinitely many nodes already in S_0 but these are not constrained at all. So one may assume that these nodes are pairwise related by the disjointness relation dc. So, in every chasing step there can be defined two disjoint sets of localities V_i^{dc} and V_i^{fin} with the following properties: for all pairwise distinct nodes in V_i^{dc} it is the case that $\mathsf{dc}(a^*, b^*) \in S_i$ and for all nodes $a^* \in V_i^{\mathsf{dc}}$ and nodes $b^* \in V_i^{fin}$ it is the case that $\mathsf{dc}(a^*, b^*) \in S_i$ and both networks are complete and relationally consistent. Now, the complete constraint network induced by V_i^{fin} is finite and is consistent with $r_1(y^*, x^*)$. This fact follows, e.g., from the patch-work property of ω-admissible theories. I use a path-consistency algorithm or some other appropriate algorithm to find a complete and consistent set induced by $V_i^{fin} \cup \{y^*\}$ that extends the networks induced by V_i^{fin} and y^*, resp. The new node y^* is related to the nodes in V_i^{dc} by dc-edges. This step does not disturb the consistency of the whole network because every composition of some basic relation with dc results in a disjunction which again contains dc. Proceeding in this way, I finally define $\bigcup_{i=0}^{\infty} S_i$ which induces a canonical model of $\mathcal{T} \cup \mathcal{A} \cup Ax_{RCC8}$.

Now I define the closure $cl_{\perp}(\mathcal{T})$ by extending the rules (1)–(6) for $cln(\mathcal{T})$ with the rules: (i) If $B_1 \sqsubseteq \exists l, l.r$ for $\mathsf{eq} \notin r \in \mathcal{T}$, then $B_1 \sqsubseteq \perp \in cl_{\perp}$; (ii) if $B_1 \sqsubseteq \exists l, l.r_1 \in \mathcal{T}$ and $B_2 \sqsubseteq \exists l, l.r_2 \in \mathcal{T}$ such that $\mathsf{eq} \notin r_1 \cap r_2$, then $B_1 \sqcap B_2 \sqsubseteq \perp$. Using $cl_{\perp}(\mathcal{T})$ one can define a FOL query that is false in $DB(\mathcal{A})$ iff $\mathcal{T} \cup \mathcal{A}$ is satisfiable.

Proof of Theorem 4.4

The proof follows the proof of Theorem 5.15 for pure DL-Lite ontologies [4]. I adapt the chase construction to account for the RCC8 relations $r \in Rel_{RCC8}$. The main observation is that the disjunctions in r can be nearly handled as if they were predicate symbols.

Let $\phi(\mathbf{x})$ be a n-ary GCQ^+-query. If $\mathcal{T} \cup \mathcal{A}$ is not satisfiable, $cert(Q, \mathcal{T} \cup \mathcal{A})$ is the set of all n-ary tuples of constants of the ontology O. But Proposition 4.2 states that satisfiability is FOL rewritable. So I can assume, that O is satisfiable.

Let pr be the UCQ resulting from applying the algorithm to $\phi(\mathbf{x})$ and O. I have to show that $cert(\phi(\mathbf{x}), O) = (pr)^{DB(\mathcal{A})}$. I proceed in two main steps. First,

Chasing Rule (R)
If $B(x) \in S_i$ and there are no y, y^*, x^* s.t. $\{R_1(x, y), l(y, y^*), l(x, x^*), r_1(y^*, x^*)\}$ is contained in S_i, then let $S_{i+1} = S_i \cup \{R_1(x, y), l(y, y^*), l(x, x^*), r_1(y^*, x^*)\}$. The constants y, y^* are completely new constants not appearing in S_i. The constant x^* is the old x^* if already in S_i, otherwise it is also a completely new constant symbol.

Fig. 4.4 Additional chasing rule that accounts for $\exists U_1, U_2.r$-concepts

we construct a chase-like set $chase^*(O)$, declare what it means to answer $\phi(\mathbf{x})$ with respect to $chase^*(O)$, resulting in the set $ans(\phi(\mathbf{x}), chase^*(O))$, and then show that $ans(\phi(\mathbf{x}), chase^*(O)) = cert(\phi(\mathbf{x}), O)$. In the second step, we show that $ans(\phi(\mathbf{x}), chase^*(O)) = (pr)^{DB(\mathcal{A})}$.

First Step. I construct a chase-like set $chase^*(O)$ that will be the basis for proving the correctness and completeness of our algorithm. I use the chase rules of Definition 2.10 and the special rule (R) of Fig. 4.4 to build $chase^*(O)$. Every time (R) is applied to yield a new abox S_i, the resulting constraint network in S_i is saturated by calculating the minimal labels between the new added region constants and the other region constants. The application of (R) does not constrain the RCC8-relations between the old regions and even the following stronger observation holds: Let (R) be applied to a tbox axiom of the form $A \sqsubseteq \exists\tilde{R}, l.r$ and $A(a) \in S_i$ resulting in the addition of $R(a, b)$, $l(b, b^*)$ and $r(b^*, a^*)$. Then it is enough to consider all $c^* \in S_i$ and all relations r_{c^*,a^*} such that $r_{c^*,a^*}(c^*, a^*) \in S_i$. The composition table gives the outcome $r_{c^*,a^*}; r = r'_{c^*,b^*}$ and one adds $r'_{c^*,b^*}(c^*, b^*)$ to S_i. This step will be called the step of triangle completion. After the triangle completion step one closes the assertions up with respect to the subset relation between RCC8-relations and with respect to symmetry. I.e., if $r_1(x^*, y^*)$ is added to S_i, then one also adds $r_2(x^*, y^*)$ for all r_2 such that $r_1 \subseteq r_2$ and $r_2^{-1}(y^*, x^*)$. For different c_1^*, c_2^*, assertions of the form $r_{c_1^*,b^*}(c_1^*, b^*)$ and $r_{c_2^*,b^*}(c_2^*, b^*)$ do not constrain each other (because of the patchwork property). As the number of regions is finite and we excluded the non-atomicity axiom, the saturation leads to a finite set S_{i+k} (for some $k \in \mathbb{N}$) that is a superset of S_i. Let $chase^*(O) = \bigcup S_i$ be the union of all aboxes constructed in this way. The set $chase^*(O)$ does not induce a single canonical model. But it is universal in the following sense:

(*) For every model $\mathfrak{I}$ of O define a model $\mathfrak{I}_c$ from $chase^*(O)$ by taking a (consistent) configuration of the contained RCC8-network and taking the minimal model of this configuration and the thematic part of $chase^*(O)$. Then $\mathfrak{I}_c$ maps homomorphically to $\mathfrak{I}$.

The claim (*) holds because $\exists U_1, U_2.r$-constructs do not appear on the left-hand side of the PI axioms; hence new information on the spatial side cannot be used during the chasing process to produce new information on the thematic part.

I explain what it means to answer a GCQ^+-query with respect to $chase^*(O)$. I transform $\phi(\mathbf{x})$ into a CQ $\tau_1(\phi(\mathbf{x}))$ where the relations $r \in Rel_{RCC8}$ are considered as predicate symbols. E.g., let $\phi(x, y) = \exists R.A(x) \wedge \exists\tilde{R}_1, l.(\mathsf{tpp} \vee \mathsf{ntpp})(y)$ and $X = \tau_1(\phi(x, y), \{\exists\tilde{R}_1, l.\mathsf{tpp}(y)\})$ for short. Then

$$X = \exists z, z^*, x^*.\exists R.A(x) \wedge R(x, z) \wedge l(z, z^*) \wedge l(x, x^*) \wedge (\mathsf{tpp} \vee \mathsf{ntpp})(z^*, x^*)$$

The set of answers $ans(chase^*(O), \phi(\mathbf{x}))$ is defined by homomorphisms of the atoms of $\tau_1(\phi(\mathbf{x}))$ into $chase^*(O)$.

$(a_1, \ldots, a_n) \in ans(chase^*(O), \phi(\mathbf{x}))$ iff there is a homomorphism h from $\tau_1(\phi(\mathbf{x}))$ into $chase^*(O)$ with $h(x_i) = a_i$ (for $i \in \{1, \ldots, n\}$). The homomorphic image of $\phi(\mathbf{x})$ in $chase^*(O)$ is called a witness of $\phi(\mathbf{x})$ w.r.t. $\mathbf{a}$ in $chase^*(O)$. Clearly, if $\mathfrak{I} \models chase^*(O)$ and $\mathbf{a} \in ans(chase^*(O), \phi(\mathbf{x}))$, then $\mathfrak{I} \models \phi(\mathbf{x}/\mathbf{a})$.

I now prove $ans(\phi(\mathbf{x}), chase^*(O)) = cert(\phi(\mathbf{x}), O)$.

$\subseteq$-direction: Let $\mathbf{a} \in ans(chase^*(O), \phi(\mathbf{x}))$. Let $\mathfrak{I} \models O$ and I_c be the model according to claim (*). Because $\mathfrak{I}_c \models \phi(\mathbf{x}/\mathbf{a})$, it follows that $\mathfrak{I} \models \phi(\mathbf{x}/\mathbf{a})$.

$\supseteq$-direction: Let $\mathbf{a} \in cert(O, \phi(\mathbf{x}))$. For every $\mathfrak{I} \models O$ consider $\mathfrak{I}_c$. All these $\mathfrak{I}_c$ differ at most on the interpretations of the RCC8-Relations which are assigned to regions x^*, y^*. Consider for all x^*, y^* the assertion $r(x^*, y^*)$, $r \in Rel_{RCC8}$ where $r_i \in r$ iff there is $\mathfrak{I}_c$ such that $\mathfrak{I}_c \models r_i(x^*, y^*)$. Then $r(x^*, y^*)$ is in $chase^*(\phi(\mathbf{x}), O)$. Therefore we find a homomorphism h from $\phi(\mathbf{x})$ onto $chase(\phi(\mathbf{x}), O)$ with $h(\mathbf{x}) = \mathbf{a}$.

Second Step. Let pr be the outcome of the algorithm applied to $\phi(\mathbf{x})$. I show $pr^{DB(\mathcal{A})} = ans(chase^*(O), \phi(\mathbf{x}))$.

$\subseteq$-direction: Let $\psi = \psi(\mathbf{x}) \in pr$ be a conjunctive n-ary query. I have to show $\psi(\mathbf{x})^{DB(\mathcal{A})} \subseteq ans(chase^*(O), \phi(\mathbf{x}))$. This can be done by induction over the number of steps that are needed in order to construct $\psi(\mathbf{x})$ in the PerfectRef algorithm. In the base case $\psi \in \tau_1(Q)$. The assertion follows directly from the fact that $DB(\mathcal{A})$ is contained in $chase^*(O)$. Inductive step. Let $\psi = \psi_{i+1}$ and ψ_{i+1} be the outcome of applying one of the steps in the algorithm to ψ_i. If the steps are those contained in the original PerfectRef-algorithm we can argue in the same line as in the proof of Lemma 5.13 of [4]. In the other cases the induction steps are provable because of the correctness of the implicit deductions.

$\supseteq$-direction: This is the direction showing the completeness of the algorithm. Let $\mathbf{a} \in ans(chase^*(O), \phi(\mathbf{x}))$. So there is a witness of $\mathbf{a}$ w.r.t. $\tau_1(\phi(\mathbf{x}))$ in $chase^*(O)$. This witness lies in some S_k of the chase $chase^*(O)$ and shall be denoted $\mathcal{G}_k$. One has to find a $\psi(\mathbf{x}) \in pr$ such that it has a witness in the abox $\mathcal{A}$. This can be proved by considering the pre-witness of $\mathbf{a}$ with respect to $\phi(\mathbf{x})$ in all S_i for $i \leq k$. The pre-witness of $\mathbf{a}$ with respect to $\psi(\mathbf{x})$ in S_i is defined by:

$$
\begin{aligned}
\mathcal{G}_i = \bigcup_{\beta' \in \mathcal{G}_k} \{\beta \in S_i \mid {} & \beta \text{ is an ancestor of } \beta' \text{ in } S_k \text{ and} \\
& \text{there exists no successor of } \beta \text{ in } S_i \text{ that is an ancestor of } \beta' \text{ in } S_k\}
\end{aligned}
$$

By induction on i ($i \in \{0, \ldots, k\}$) one can find a $\psi(\mathbf{x}) \in pr$ such that the pre-witness of $\mathbf{a}$ with respect to $\phi(\mathbf{x})$ in S_{k-i} is a witness for $\psi(\mathbf{x})$. By induction assumption there is $\psi'(\mathbf{x}) \in pr$ such that $\mathcal{G}_{k-i+1}$ is a witness of $\mathbf{a}$ w.r.t. $\psi'(\mathbf{x})'$ in S_{k-i+1}. If S_{k-i+1} results from S_{k-i} by application of one of the chase rules in Definition 2.10, then the argument proceeds in the same manner as in the proof of Lemma 5.13 in the paper of [4]. Otherwise, S_{k-i+1} is constructed by applying rule (R) or one of the saturation steps (triangle completion, upward closure, symmetry closure, resp.) following the application of rule (R). But all these steps have a corresponding case in the algorithm.

Proof of Lemma 4.6

According to a result of Goldreich the reachability problem in symmetric (undirected) graphs is logspace complete—where graph reachability asks whether for nodes s, t in G there is a path between s and t. By reducing this problem to the satisfiability test for RCC3 networks I will have shown that the latter problem is LOGSPACE hard itself. So let be given a (symmetric) graph $G = (V, E)$ and nodes $s, t \in V$. I define the network N in the following way (see Fig. 4.5): Let $V = \{v_1, \ldots, v_n\}$ be an enumeration of the nodes of G. Without loss of generality let $s = v_1$ and $t = v_n$ and let $\mathcal{B} = \mathcal{B}_{RCC3}$. Nodes of N are given by $V \cup \{a\}$ where $a \notin V$. Labelled edges of N are given by: $s\{\mathsf{DR}\}a$; $t\{\mathsf{ONE}\}a$; $v_i\{\mathcal{B}\}a$ for all $i \neq 1, n$; $v_i\{\mathsf{EQ}\}v_j$ if $E(v_i, v_j)$; $v_i\{\mathcal{B}\}v_j$ if $\neg E(v_i, v_j)$. Now we show that the network N is satisfiable iff s and t are

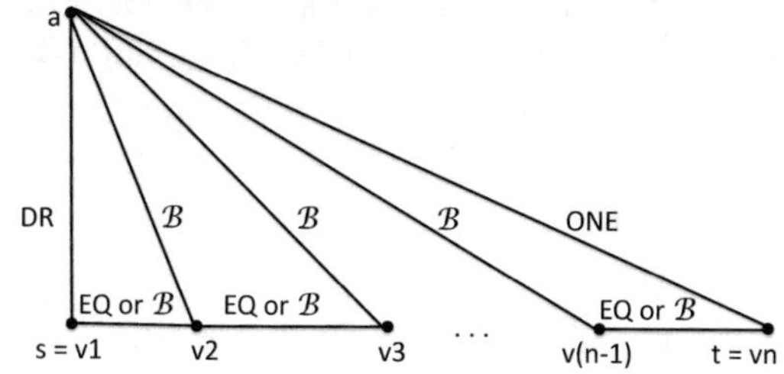

Fig. 4.5 Network N used in proof of Lemma 4.6

connected in G. Assume that s and t are connected. Then there is an EQ-path in N between them, hence $s\{\mathsf{EQ}\}t$ follows. But this contradicts $s\{\mathsf{DR}\}a$ and $t\{\mathsf{ONE}\}a$. Now assume that s and t are not connected; then there is no path consisting only of EQ-labels between s and t. The graph G consists of at least 2 components, and s, t are in different components. We define a consistent configuration as follows: For all nodes v, v' in the component in which s is contained, let $v\{\mathsf{DR}\}a$ and $v\{\mathsf{EQ}\}v'$. For all nodes v, v' in the component of t let $v\{\mathsf{ONE}\}a$ and $v\{\mathsf{EQ}\}v'$. For all nodes v, v' in the other components let $v\{\mathsf{DR}\}a$ and $v\{\mathsf{EQ}\}v'$. For all nodes v, v' which have not a label yet, let $v\{\mathsf{DR}\}v'$. Two remarks : 1. EQ-edges for edges $E(v_i, v_j)$ in G with $j > i + 1$ are not shown in Fig. 4.5. 2. I inserted edges labelled $\mathcal{B}$ for better illustrations. But these are not needed.

Proof of Proposition 4.7

I construct a generic tbox $\mathcal{T}_g$ that allows one to encode any RCC3 constraint network so that checking the consistency of RCC3 constraint networks is reducible to a satisfiability check of this tbox and a spatially complete abox. Let for every $r \in Rel_{RCC3}$ be given role symbols R_r^1, R_r^2. The generic tbox $\mathcal{T}_g$ has for every $r \in Rel_{RCC3}$

a concept symbol A_r and a corresponding axiom with the content that all instances of A_r have paths over the abstract features R_1 resp. R_2 to regions that are r-related.

$$\mathcal{T}_g = \{A_r \sqsubseteq \exists \tilde{R}^1_r, \tilde{R}^2_r.r, (\text{funct } l, R^1_r, R^2_r) \mid r \in Rel_{RCC3}\} \tag{4.1}$$

Now, let N be an arbitrary RCC3 constraint network which has to be tested for relational consistency. Let $\mathcal{A}_N$ be an abox such that for every $r(a, b)$ in N three new constants are introduced: x_{ab}, x_a, x_b.

$$\mathcal{A}_N = \{A_r(x_{ab}), R^1_r(x_{ab}, x_a), R^2_r(x_{ab}, x_b) \mid r(a, b) \in N\} \tag{4.2}$$

The construction entails: $\mathcal{T}_g \cup \mathcal{A}_N \cup Ax_{RCC3}$ is satisfiable iff $N \cup Ax_{RCC3}$ is satisfiable. If the data complexity of the satisfiability check for DL-Lite$^{\sqcap,+}_{\mathcal{F},\mathcal{R}}$(RCC3)-ontologies were in AC^0, then the consistency of constraint networks could be tested in AC^0, too. (Note that $\mathcal{T}_g$ is a fixed tbox.) But checking the consistency of RCC3 constraint networks is LOGSPACE-hard and $AC^0 \subsetneq$ LOGSPACE.

Proof of Proposition 4.10

Sufficiency of the condition for unsatisfiability is clear. The proof for necessity is done by induction on the number of =-marked labels. So let be given a RCC$^=$ network N. One may assume, that the network has for every pair of nodes exactly one labelled edge between. I assume further that the edges are undirected as all relations are symmetric.

Base case: Assume that none of the four conditions hold. As there are no marked labels, then N unsatisfiability can occur only, if N contains $\mathsf{DR}(v, v)$ for some node v (first condition) or $\mathsf{DR}(v, w)$ and $\mathsf{O}(v, w)$ (fourth condition). But these cases are excluded by assumption.

Induction step: Let N contain n marked labelled edges and assume that for all networks N' with $n - 1$ marked labelled edges unsatisfiability of N' implies one of the conditions. Now, assume that for N no one of the four conditions holds. We have to show that N is satisfiable. Take an arbitrary =-marked labelled edge between nodes v, w. The label λ of this edge is either $\mathsf{O}^=$ or $\mathcal{B}^=$. We define a new network N' which results as a contraction from N by identifying v and w to the node z. For N' we may assume again that it contains for every pair of nodes exactly one labelled edge: If in N we have $r_1(v, k)$ and $r_2(w, k)$, then the edge between z and k in N' results as $r_1 \cap r_2$ and is =-marked iff r_1 or r_2 is marked. Clearly $r_1 \cap r_2$ is not empty, as otherwise one would have a contradiction to the fact that N does not fulfil conditions 3 and 4. Clearly N is satisfiable iff N' is satisfiable. Assume to the contrary that N' is unsatisfiable. Hence, one of the four conditions holds for N': 1) Assume N' contains $\mathsf{DR}(v', v')$ or $\mathsf{DR}^=(v', v')$ for some node v'. If v' is not z, then also $\mathsf{DR}(v', v') \in N$ resp. $\mathsf{DR}^=(v', v') \in N$—contradicting the fact that N does not fulfil condition 1. Otherwise $v' = z$, but this cannot be the case either, as there are no self-loops $\mathsf{DR}(v, v)$, $\mathsf{DR}^{=}(v, v)$, $\mathsf{DR}(w, w)$, $\mathsf{DR}^{=}(w, w)$ in N nor $\mathsf{DR}(v, w)$ or $\mathsf{DR}(v, w)$. 2) Assume N' contains $\mathsf{DR}^{=}(v', w')$. If neither v' nor w' is z, this contradicts the fact that N does not fulfil the second condition. Otherwise $v' = w' = z$, leading to a contradiction with the fact that N does not fulfil the first condition. 3) Assume N' contains a cycle in which there is $\mathsf{DR}(v', w')$ and there is a path from v' to w' such that every label on the path is $\mathcal{B}^{=}$ or $\mathsf{O}^{=}$. The case that the path does not contain z immediately leads to a contradiction. Otherwise, the path extends to a path in N fulfilling the 3. condition—contradiction. Similarly, if N' fulfils condition 4, the verifying path can be extended to a path in N fulfilling condition 4—contradiction.

Proof of Proposition 4.11

The idea is to find concepts that are unsatisfiable according to the tbox (this amounts to constructing the negative closure as in the proofs for pure DL-Lite [4, Def. 4.7, p. 292]). These are formulated as boolean queries, and the (finite) disjunctions of these queries are answered over the abox. For example, if $A_1 \sqsubseteq \neg A_2$ is in the tbox, then the query contains $\exists x(A_1(x) \wedge A_2(x))$ as disjunct. The introduction of concepts of the form $\exists U_1, U_2.r$ enlarges the potential conflicts of a tbox $\mathcal{T}$ with an abox $\mathcal{A}$. So in addition to the FOL queries that result from the negative closure of the tbox, one has to find queries for the potential conflicts in the four conditions of Proposition 4.10. The resulting conjunctive queries have further to be fed into a perfect rewriting algorithm like PerfectRef for pure DL-Lite [4] in order to capture the implications of the tbox.

Concerning the first two conditions one has to deal with axioms of the form $B \sqsubseteq \exists l, l.\mathsf{DR}$. If this axiom occurs in the tbox $\mathcal{T}$, then one has to produce the CQ $\exists x.B(x)$. Similarly, if $\{B \sqsubseteq \exists \tilde{R}, \tilde{R}.\mathsf{DR}, (\text{funct } R)\} \subseteq \mathcal{T}$, then the CQ $\exists x.B(x)$ has to be added. Also if $\{B \sqsubseteq \exists \tilde{R}_1, \tilde{R}_2.\mathsf{DR}, (\text{funct } R_1, \text{funct } R_2)\} \subseteq \mathcal{T}$, then one may get a conflict of the first kind and hence one has to add a CQ $\exists x, y[B(x) \wedge R_1(x, y) \wedge R_2(x, y)]$. Concerning the third and fourth condition I note that this can be only the case if the =-marked paths have maximal length one. Otherwise one already has a contradiction of the abox with the functionality assertions in the tbox. Hence, one considers only the case of pairs of tbox axioms of the general form $A \sqsubseteq \exists \tilde{R}_1, \tilde{R}_2.\mathsf{DR}$ and $B \sqsubseteq \exists \tilde{R}_3, \tilde{R}_4.\mathsf{O}$ with (funct R_1, R_2, R_3, R_4). In this case one can feed into the PerfectRef algorithm also the CQ $\exists x, y, z, w[A(x) \wedge B(y) \wedge R_1(x, z) \wedge R_3(y, z) \wedge R_2(x, w) \wedge R_4(y, w)]$.

Chapter 5
Representation Theorems for Stream Processing

Abstract This chapter outlines a formal and foundational treatment of stream processing from the word perspective where stream processing and stream queries are modelled as functions mapping (finite or infinite) words to words. The main relevant aspects of streams—potential infinity, ubiquity, and ordering/recency—are taken into account in order to give an axiomatic characterisation of various natural classes of stream functions.

5.1 Introduction

Stream processing has been and is still a highly relevant research topic in CS. There are quite a few research paper titles hinting concisely to various important aspects of stream processing, be it the ubiquity of streams due to the temporality of most data ("It's a streaming world!", [13]), or its potential infinity in contrast to data stored in a static database system ("Streams are forever", [15]), or the importance of the order in which data are streamed ("Order matters", [41]).

These aspects are relevant for all levels of stream processing, in particular for classical stream processing on the sensor-data level, e.g., within sensor networks or for agent reasoning percepts, or on the relational data level, e.g., within stream data management systems. Recent interest on high-level declarative stream processing w.r.t. a terminological knowledge base (ontology) have lead to additional aspects becoming relevant. In this scenario, the enduser accesses all possibly heterogeneous data sources (static and streaming) via a declarative query language using the signature of the ontology. Some recent projects, such as BOEMIE[1] and CASAM[2], demonstrated how such a uniform ontology interface could be used to realise (abductive) interpretation of multimedia streaming data, which combine video streams, audio streams, and text streams with annotations [20]. This kind of convenience

[1] http://cordis.europa.eu/ist/kct/boemie_synopsis.htm

[2] http://cordis.europa.eu/project/rcn/85475_en.html

Ö. L. Özçep, *Representation Theorems in Computer Science*,
https://doi.org/10.1007/978-3-030-25785-9_5

and flexibility for the end-user leads to challenging aspects for the streaming engine which has to provide the homogeneous ontology view over the data and which has to guarantee that all (and only) answers w.r.t. the ontology are captured.

Stream processing has strong connections to the processing of temporal data in a temporal DB using temporal (logic) operators. Nonetheless, the scenarios, the objects of interest, the theoretical and practical challenges are different from those of temporal logics/temporal DBs. While query answering on temporal DBs is a one-step activity on static historical data, answering queries on streams is a reactive, continuous activity (hence the notion of a continuous query).

The main challenging fact of stream processing is the potential infinity of the data. In particular, potential infinity means that one cannot apply a one-shot query-answering procedure and, moreover, that one should be able to process the data in one-pass style (consuming small space) as one is not going to store all the data. On top of these theoretically fundamental challenges one has to deal in practice with high-pace stream data where the velocity at which the stream data arrive may not be even constant (burstiness). Furthermore, in contrast to temporal DBs settings, in stream scenarios usually more than one query is registered to the stream engine. The reason is, first, that there are usually many stream sources corresponding, e.g., to different sensors and control units, and, second, that there are many features to be monitored, e.g., various statistical and time-series features or various event patterns. The focus in this chapter is on the potential infinity and the one-pass processing constraint. I also consider the multiple-stream aspect (see Sect. 5.6), but rather from the theoretical perspective of correct semantics, neglecting aspects of optimisation.

In this chapter I outline a formal and foundational treatment of stream processing from the (infinite) word perspective along the line of [22] in which all the main aspects of streams mentioned above (potential infinity, ubiquity, and ordering/recency) are taken into account. The main results of these considerations, which are relevant for all levels of stream processing, are representation results of the first category, namely axiomatic characterisations of various stream query classes. In particular, the whole class of genuine stream queries is shown to be representable as stream queries induced by window (alias kernel) functions—justifying formally that many papers on stream processing deal with the window constructor as a means to cope with the potential infinity of streams. Subclasses of these (with window range and slide) can also be characterised with specific axioms expressing factorisation properties.

In the following chapter, the focus is on high-level data stream processing as required, say, for high-level streams (belief of states, planned actions etc.) in rational agents acting in a dynamic environment. As an example of a query language for high-level declarative stream processing I describe the stream-temporal query framework (STARQL), which was developed in the EU project Optique[3].

The chapter is structured as follows: Section 5.2 gives the necessary terminology for handling streams as finite or infinite words. The first representation result of this chapter is contained in Sect. 5.3: prefix-determined stream queries can be described by windows. Results on characterizing stream queries induced by constant-width

[3] http://optique-project.eu/

windows with factorisation properties are given in Sect. 5.4, results on adding time in the word model can be found in Sect. 5.5, and results on building bounded-memory windows are developed in Sect. 5.6. The chapter closes with a section on related work (Sect. 5.7) and a summary (Sect. 5.8).

Contents of this chapter are published in [32] and [31].

5.2 Preliminaries

Though there are various stream definitions over various research communities, and even within researchers of the same community, a common aspect of all streams is that they are constituted by a potentially infinite sequence of data elements from some domain. The sequence is assumed to be ordered-isomorphic to the natural numbers, so that there is always a least element of a subset of the stream (and additionally there is always a unique predecessor and successor of an element in the stream). In this chapter it will be convenient to work with the following very basic definition of streams based on words over an alphabet D.

Definition 5.1 A *stream* is a finite or infinite sequence of elements from a domain D. Formally, the set of *finite streams* is the set of finite words D^* over the alphabet D. The set of *infinite streams* is the set of ω-words D^ω over D. The set of (all) streams is denoted $D^\infty = D^* \cup D^\omega$.

$D^{\leq n}$ denotes the set of words over D with length less than or equal to n. For any finite stream s the length of s is denoted by $|s|$. For infinite streams s let $|s| = \infty$ for some fixed object $\infty \notin \mathbb{N}$. For $n \in \mathbb{N}$ with $1 \leq n \leq |s|$, let $s^{=n}$ be the n-th element in the stream s. For $n = 0$ let $s^{=n} = \epsilon =$ the empty word. $s^{\leq n}$ denotes the n-prefix of s, $s^{\geq n}$ is the suffix of s such that $s^{\leq n-1} \circ s^{\geq n} = s$. For an interval $[j, k]$ for $1 \leq j \leq k$, $s^{[j,k]}$ is the stream of elements of s such that $s = s^{\leq j-1} \circ s^{[j,k]} \circ s^{\geq k+1} = s$. For the special instance of the interval operator $s^{[|s|-n,|s|]}$ I also use the notation $last_n(s)$ as it denotes the word consisting of the last n symbols of s, i.e., the *n-suffix*, which is defined for finite s only. For a finite stream $w \in D^*$ and a set of streams X, $w \circ X$ or shorter wX is defined as the set $\{s \in D^\infty \mid \text{There is } s' \in X \text{ s.t. } s = w \circ s'\}$. That s is a *prefix* of s', for short: $s \sqsubseteq s'$, means that there is a word y such that $s' = s \circ y$. If s is a prefix of s', then $s' -_{\sqsubseteq} s$ is the word y such that $s' = s \circ y$. If all letters of s occur in s' in the ordering of s (but perhaps not directly next to each other) then s is called a *subsequence* of s'. If $s' = usv$ for $u \in D^*$ and $v \in D^\infty$, then s is called a *subword* of s'.

Sometimes I am going to write streams in the word notation, sometimes writing out concatenation $\circ$ explicitly. In the next chapter I will follow the convention of the RDF Stream processing Community[4] and represent streams in set notation, it being understood that there is an ordering isomorphic to the ordering of the natural numbers. In this case I then use $\leq_{ar}$ to denote the ordering between the elements of the streams. The acronym *ar* should remind the reader of the "arrival" ordering.

[4] `https://www.w3.org/community/rsp/`

I should stress that with the above definition of "stream" I am abstracting from the underlying stream engine that processes stepwise the arriving data elements of the stream. The engine does not actually see the stream in the above sense but only sees some initial part of it and expects a possibly new element to arrive. So, the concept defined above under the term "stream" would be better termed "stream trace". But again, I follow the terminological practice in the stream community.

The basic definition of streams above is general enough to capture all of the different forms of streams considered in this and the next chapter. Amongst these is the important category of temporal streams. Though not restricted to, temporal streams are one of the most common stream types to occur in applications. In this case the domain D consists of pairs of objects (d, t) were d is an element of a set, which is termed *object domain* and an element t from a *time domain* $(T, \leq_T)$. Here, $\leq_T$ is a binary relation on the set T. Depending on the kind of application, T may consist of time points, intervals or even multidimensional time entities. In many cases $\leq_T$ obeys further constraints that justify the symbol $\leq$ for an order relation. In particular, I am going to consider so-called *flows of time* $(T, \leq_T)$ where T is a set of time points and $\leq_T$ is a linear, transitive, reflexive, and antisymmetric relation $\leq$ on T. And even more constrained, in the examples introduced below, I work with a discrete time flow T, e.g., natural numbers with the usual ordering. But note that the query language STARQL introduced in the next chapter is also applicable to temporal streams with dense time domains such as the rational numbers $\mathbb{Q}$ or even continuous time domains such as the real number $\mathbb{R}$. The only additional constraint is that for windows based on the flow of time (and not the number of arriving elements) one has to ensure that there are no cumulation points. Together with the synchronicity assumption (see below) this constraint says:

(No Cumulation) The sequence of time annotations t_i in a stream is finite (if the stream is finite) or is monotonically increasing without any cumulation points (i.e., there is no $t \in T$ such that for any ϵ infinitely many t_i are in ϵ distance from t, $\epsilon > 0$).

According to the definition of a temporal stream, it may happen that two different objects d_1 and d_2 with the same timestamp t occur, and even many different occurrences of the same time tagged tuple $d\langle t\rangle$ may occur. Moreover, it may be the case that there are timestamps which do not occur in the stream. The latter is useful in particular for those situations where there is no information at some time point from T—and hence this representation is also useful to model varying frequencies of stream element occurrences.

In a *synchronised stream setting*, one demands that the timestamps in the arrival ordering make up a monotonically increasing sequence, so that $\leq_T$ is conform with $\leq_{ar}$. In an *asynchronous stream setting*, it may be the case that elements with earlier timestamps arrive later than data with later timestamps. In particular, in this latter case, the order needed for a stream to be isomorphic to the natural numbers does not have to adhere to the order $\leq_T$ of the flow of time $(T, \leq_T)$.

The distinction regarding synchronicity hints to possible layers of streams. For example, in [27], the authors distinguish between raw, physical, and logical streams.

Raw streams correspond to temporal streams according to our definition and are intended to model the stream data arriving at a data stream management system (DSMS). Logical streams are abstractions of raw streams where the order of the sequence is ignored, so they can be defined as multi-sets of timestamped tuples.[5] Physical streams allow not only for timestamps but also half-open intervals $[t_s, t_e)$ as time annotations, the ordering of the stream being non-decreasing w.r.t. t_s.

Orthogonally to these layered distinction of stream types, streams are categorised according to the type of the domains. In the context of ontology-based data access for streams—as discussed in the following chapter—at least two different domains D of streamed objects have to be taken into consideration. The first domain consists of relational tuples from some schema. In this case, I call the stream a *relational* stream. The second domain is made up by data facts, either represented as abox assertions or as RDF triples. In this case I just talk of a stream of abox assertions, RDF streams resp. In case of relational streams all tuples adhere to the same schema, hence relational streams are homogeneous, whilst in the case of streams of abox assertions/RDF triples the used logical vocabulary is not restricted to a specific relation symbol. In so far, these streams are inhomogeneous. Nonetheless, one may restrict the signature of the assertions, thereby replacing the role of the relational schema in relational streams by a signature.

5.3 Stream Queries in the Word Perspective

This and the following sections are on the foundations of streams, discussing the paper by Gurevich and colleagues [22] and deriving on top of it representation results. The main result of this section is that genuine stream queries are window-based.

The main aim is to axiomatically characterise different classes of functions of the form $Q : D_1^{\infty} \longrightarrow D_2^{\infty}$. So the focus is on total functions which map a finite or infinite stream over a given domain D_1 to a finite or infinite stream on a domain D_2. In this section I am going to assume without loss of generality the same domain $D = D_1 = D_2$ for inputs and outputs. All functions of this form will be denoted by Q or primed and indexed variants—and generally I use the term "stream **q**uery" to denote these functions from streams to streams. This is analogous to the use of "query" in finite model theory (see also Chap. 2). In the next chapter, I also consider declarative stream query languages, in particular the stream query language STARQL, which induces queries in the sense used in the current chapter.

By considering the union of finite and infinite streams as potential domains and ranges of stream queries I am following the approach of [22]. Later I also consider—thereby following [42]—functions where the domain (resp. the range) is either the set of finite streams D^* only or the set of infinite streams D^{ω} only. The more general definition of a stream query according to [22] allows to cover different scenarios that

[5] Note that the original definition in [27] would also consider uncountable sets as streams, if one chooses $\mathbb{R}$ as time domain, so that the intuition of a stream as a set ordered as the natural numbers cannot be applied here.

have found interest in CS. In particular, allowing for finite streams as potential inputs of stream queries allows to cover scenarios where a stream engine is allowed to be stopped after a finite number of steps. It also covers the case where the engine gets informed about the fact that the input stream is finite so that it can stop processing because there is no element to come anymore. A framework for implementing such "informed" stream processing is developed under the term "punctuation semantics" in [40].

Allowing for finite streams in the output covers scenarios where one is interested in recognising (infinite) streams in the same way one recognises, say, regular languages by finite automata: in the special case where the range of a stream query is $D = \{0, 1\}$, one can define those input streams to be accepted that are mapped to 1, the others, i.e., those mapped to 0, to be not accepted.

So with this general definition one captures generally two categories of stream processing, *local stream processing* and *global stream processing*. In the local approach (such as monitoring) one is interested only in the initial part up to the current time point and applies some local operation (such as averaging) on this initial part or even some finite suffix of this initial part (see window operators below). In the global approach some property of the whole stream is learned by processing successively growing initial parts. Examples of the former kind will be dealt extensively in the next section on STARQL. An example of the latter kind is given in formal learning theory. In [17], e.g., an ontology is learned from a sequence of assertions.

In the framework of [22], multiple streams are handled by allowing the domain elements to be tagged with provenance information, in particular information on the stream source from which the data element originates. Using these tags it is possible to formulate queries such as INTERSECT which asks whether there is a same element in two input streams up to some time n, in which case it outputs a 1 and otherwise a 0. So with this approach stream queries on interleaved stream queries can be modelled. But there is no control on the interleaving as in state-of-the art stream query languages following a pipeline architecture. For this, the framework would have to be extended to handle functions of the form $Q : D^{\infty} \times \cdots \times D^{\infty} \longrightarrow D^{\infty}$. In Sect. 5.6 I propose such an extension in the context of bounded-memory queries.

With the formal framework for modelling streams as words and for modelling stream processing by word-to-word functions we have the basic language to state first representation theorem. For this I first describe axiomatically the formal properties that any stream query is expected to fulfil. The first observation is that, though any function $Q : D^{\infty} \longrightarrow D^{\infty}$ is termed stream operator, only those operators that produce their outputs successively by considering one input-stream element after the other, have to be accepted as genuine stream queries. This intuition on the continuous, successive production of the output can be formalised by saying the output is determined only by finite prefixes. This is the content of the following axiom denoted (FP$^{\infty}$).[6]

(FP$^{\infty}$) For all $s \in D^{\infty}$ and all $u \in D^*$: if $Q(s) \in uD^{\infty}$, then there is a $w \in D^*$ such that $s \in wD^{\infty} \subseteq Q^{-1}(uD^{\infty})$.

[6] (FP$^{\infty}$) is my generalization of the axiom (FP) in [42].

As a further aspect of stream processing one can consider data-driveness: the actions of the engine processing the stream (to be read as: the outputs produced by the function Q in this abstract setting) are triggered by the input stream not by the engine itself. In particular this means that finite streams are allowed to be mapped only to finite streams. I fix this condition as further axiom:

(F2F) For all $s \in D^*$ it holds that: $Q(s) \in D^*$.

The dual constraint, which is going to be used later, is that infinite streams must be mapped to infinite streams. Accepting this axiom means that one considers, using the terminology introduced above, local stream processing and not global stream processing.

(I2I) For all $s \in D^\omega$ it holds that $Q(s) \in D^\omega$.

Already axioms (FP$^\infty$) and (F2F) can be represented by a class of operators based on the notion of abstract computability according to [22]. The very general notion of an *abstract computable* stream function is that of a function which is incrementally computed by calculations of finite prefixes of the stream w.r.t. a function called *kernel*. More concretely, let $K : D^* \longrightarrow D^*$ be a function from finite words to finite words. Then define the *stream query* $Repeat(K) : D^\infty \longrightarrow D^\infty$ *induced by kernel* K as

$$Repeat(K) : s \mapsto \bigcirc_{j=0}^{|s|} K(s^{\leq j})$$

Definition 5.2 A query Q is *abstract computable* (AC) iff there is a kernel such that $Q(s) = Repeat(K)(s)$. We denote by $\mathcal{QL}_{AC}$ the set of AC queries.

Using a more familiar speak from the stream processing community, the kernel operator is a *window operator*, more concretely, an unbounded window operator. I will stick to the window terminology in the following.

An important aspect for the usability of queries is that queries can be cascaded, or—formally—that they can be composed. This is indeed the case for AC (and also for SAC queries, see below) as shown in [22].

The definition of abstract computable functions is quite general and, in fact, Definition 5.2 does not say anything about the computability of kernels K. The main idea for this definition stems from the theory of computability on real numbers, termed “theory of type-2 effectivity” as outlined in [42]. The robustness or—say—usability of this notion is shown by the fact that abstract computable queries represent all functions that map finite streams to finite streams, i.e. fulfil (F2F), and that are finite prefix determined, i.e. fulfil (FP$^\infty$). This is the first representation result of this chapter.

Theorem 5.3 *Abstract computable functions represent the class of stream queries fulfilling (F2F) and (FP$^\infty$).*

This theorem underlines the importance of the window concept because it shows that any stream query implementing the core idea of an incremental, continuous operation on the ever growing prefix of incoming stream elements can be represented as a window-based function. So it is no surprise that most of the stream literature discusses, in one or other form, window functions.

Theorem 5.3 corresponds exactly to Theorem 9 of [22] with the only difference that the authors talk of continuity instead of the Postulate (FP$^\infty$). For sake of completeness I give the definition of continuity according to [22] and, in the appendix, the proof of Theorem 5.3. The topology for which continuity is declared is similar to but not exactly the same as the topology induced by the Cantor metric for infinite streams.[7] Let $p \in D^*$. The *open ball* $B(p)$ around p is defined as the set of all streams (finite or infinite) having p as prefix:

$$B(p) = \{s \in D^\infty \mid p \sqsubseteq s\} = pD^\infty$$

A query $Q : D^\infty \longrightarrow D^\infty$ is called *continuous* [22] iff for every open ball B the pre-image $Q^{-1}(B)$ is a union of (possibly infinitely many) open balls. It is easy to see that functions continuous w.r.t. this topology are exactly those functions that fulfil FP^∞: An open ball B has the form $B = B(u) = uD^\infty$ for some $u \in D^*$. Continuity means that $Q^{-1}(uD^\infty)$ is open, i.e., that around any s with $Q(s) \in uD^\infty$ one can find an open ball $wD^\infty \ni s$.

As already noted by the authors of [22] this topology is not metrisable (i.e., there is no metric inducing this topology) as the topology is not Hausdorff: a finite prefix p and an infinite extension $D^\omega \ni s \sqsupseteq p$ are not separable.

That the window abstraction can be seen as a core abstraction of stream processing on finite and infinite streams has been justified by the characterisation in Theorem 5.3. This still should be true when considering stream queries $Q : D^\omega \longrightarrow D^\omega$ which are defined only on infinite streams and that output only infinite streams. The finite prefix statement then has the following form:

(FP$^\omega$) For all $s \in D^\omega$ and all $u \in D^*$: if $Q(s) \in uD^\omega$, then there is a $w \in D^*$ such that $s \in wD^\omega \subseteq Q^{-1}(uD^\omega)$.

Clearly, any operator $Q : D^\omega \longrightarrow D^\omega$ that is generated as $Repeat(K)$ for a window function K fulfils (FP$^\omega$). But does the converse hold, too? At least this is not obvious from Theorem 5.3 as a closer inspection of the proof shows that the window K is defined by relying on the definition of Q on *finite streams*, which—in the case we consider now—is not possible because Q is defined only on infinite streams. But indeed, using a different window construction also yields the desired representation theorem.

Theorem 5.4 *Abstract computable functions of the form* $Q : D^\omega \longrightarrow D^\omega$ *represent the class of stream queries fulfilling (FP$^\omega$).*

Proof See p. 124. □

[7] The Cantor metric ρ is defined as follows: For any infinite streams $s, s' \in D^\omega$

$$\rho(s, s') = \begin{cases} 0 & \text{if s = s'} \\ \frac{1}{2^{-n}} & ifs \neq s' \text{ and } n = min\{i \mid s^{=i} \neq s'^{=i}\} \end{cases}$$

5.4 Constant-Size Windows

A simple fact following from Theorem 5.3 (and actually used in the proof of Theorem 9 of [22]) is that if Q maps finite streams to finite streams and is continuous, then, for all finite streams s and letters u, $Q(s) \sqsubseteq Q(su)$ holds. Actually, this can be easily extended to a monotonicity statement. A stream query Q is *monotone* iff it fulfils the following axiom:

(Mon) For all finite streams $s' \in D^*$ and all (finite and infinite) streams $s \in D^\infty$: if $s' \sqsubseteq s$, then $Q(s') \sqsubseteq Q(s)$.

So we get

Proposition 5.5 *(FP^∞) and (F2F) together entail (Mon).*

A stricter axiom than (Mon) is (Distribution). It actually characterises queries that are completely determined by their outputs for finite streams of length 1.

(Distribution) For all $s \in D^*$ with $|s| \geq 1$: $Q(s) \in D^*$ and for all $s' \in D^\infty$ with $|s'| \geq 1$: $Q(s \circ s') = Q(s) \circ Q(s')$.

Proposition 5.6 *Any stream query Q fulfilling (Distribution) fulfils*

$$\textit{for all } s \in D^\infty\textit{: } Q(s) = \bigcirc_{i=1}^{|s|} Q(s^{=i})$$

In particular such stream queries are AC and hence fulfil (FP^∞).

Proof See p. 126. □

In particular distributive queries are monotone:

Proposition 5.7 *(Distribution) entails (Mon) and (F2F).*

Note that the distribution axiom requires s and s' to be non-empty streams. The reason is that otherwise the stream query would have to map the empty stream onto itself: $Q(\epsilon \circ \epsilon) = Q(\epsilon) \circ Q(\epsilon)$ entails that $Q(\epsilon) = \epsilon$.

As the proof of Proposition 5.6 shows distributive stream queries can be generated by windows which depend only on the 1-suffix of a word.

For later use I define the more general notion of an n-window ($n \in \mathbb{N}$) which corresponds to the notion of a finite window of width n.

Definition 5.8 A function $K : D^* \longrightarrow Y$ that is determined by the n-suffixes, i.e., that fulfils for all words $w, u \in D^*$ with $|w| = n$ the condition

$$K(uw) = K(w)$$

is called an *n-window (function)*. If additionally $K(s) = \epsilon$, for all s with $|s| < n$, then K is called a *normal* n-window.

Stream queries generated by an n-window for some $n \in \mathbb{N}$ are called *n-window abstract computable* stream queries, for short *n-WAC* queries. The union $WAC = \bigcup_{n \in \mathbb{N}} n - WAC$ is the set of window abstract computable stream queries.

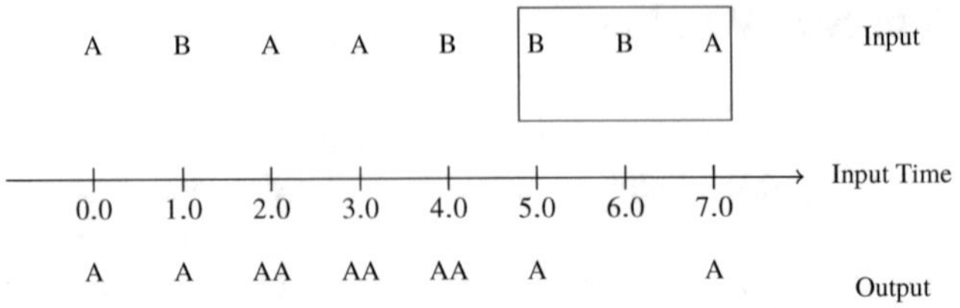

Fig. 5.1 Example of a stream query based on a constant-width window

Please note that I chose to attach the finitude property to the window. An alternative way which seems to fit better the usual practices in the steaming community is to attach the finitude to the *Repeat* functional. More concretely, one can define a new functional $Repeat^n$ which depends on the window size n as a parameter and which is defined as follows:

$$Repeat^n(K) : s \mapsto \bigcirc_{j=0}^{|s|} K(s^{[max(\{0, j-n\}), j]})$$

The reason why I chose to stick to the old *Repeat* functional is that this enables re-using the results on AC functions. For example, though it may seem trivial, with this approach it easily follows that all WAC queries are continuous.

Example 5.9 Figure 5.1 illustrates a simple query based on a 3-window K that outputs all As of the last 3 time units. The figure shows the position of the window after 7 time units where the content of the window is the word BBA. Hence the output has an A at time unit 7. All in all the input stream starting with the word $ABAABBBA$ produces an output stream starting with the word $AAAAAAAAA$. At time point 6.0 the window content is BBB, so the empty word is written to the output. Note that the time axis denotes input time. We have written the output depending on the input time. In fact, in the infinite word model of [22] time is identified by the positions of the characters in the words. This means, e.g., that at input time 2.0 the output is already at time point 4.0. See Sect. 5.5 for a further discussion of this time notion.

The following proposition gives some simple properties w.r.t. (Distribution).

Proposition 5.10

1. *0-windows correspond exactly to the constant functions on D^* with $K(u) = K(\epsilon)$ for all $w \in D^*$.*
2. *Every i-window with $i \leq j$ is also a j-window.*
3. *Stream queries fulfilling (Distribution) can be generated by a 1-window. Conversely, if a stream query is generated by a 1-window with $K(\epsilon) = \epsilon$, then it fulfils (Distribution).*

Proof See p. 126. □

A property for stream queries related to (DISTRIBUTION) is the *filter property*. Roughly, the filter property states that Q filters out elements from the input stream thereby outputting a subsequence. Remember that a subsequence w of a stream s is a word whose elements appear in s in the order of w but perhaps not in a contiguous way. The following axiom gives a possible instantiation of this property:

(FILTER)
1. $Q(\epsilon) = \epsilon$
2. For all $s \in D^{\infty}, u \in D$: $Q(us) = u \circ Q(s)$ or $Q(us) = Q(s)$.

The following proposition follows immediately.

Proposition 5.11 *All operators fulfilling (FILTER) are AC.*

Proof Define $K(\epsilon) = \epsilon$ and $K(uw) = u$ if $Q(uw) = u \circ Q(w)$ and $K(uw) = \epsilon$ if $Q(uw) = Q(w)$ for all $w \in D^*$ and $u \in D$. □

Note that the notion of a filter according to (FILTER) lies between two other filter notions. There is the most general one which requires only that for any $s \in D^{\infty}$ the output $Q(s)$ is a subsequence of s. This filter notion neglects the incremental aspect of stream processing and hence I do not investigate its properties here.[8]

A more strict version of (FILTER) can be termed time-invariant filters, abbreviated by (TI-FILTER). According to this axiom, the decision whether to incorporate the element u into the output stream does not depend on the position of u in the input stream, in other words, the filter is time invariant.

(TI-FILTER)
1. $Q(\epsilon) = \epsilon$
2. for all $s \in D^{\infty}, u \in D, w \in D^*$:
   either $Q(wus) = Q(w) \circ u \circ Q(s)$ or $Q(wus) = Q(w) \circ Q(s)$.

Clearly, all time-invariant filter queries are also distributive.

Proposition 5.12 *(TI-FILTER) entails (DISTRIBUTION)*

Proof See p. 127. □

There are even further different notions of filters coming from the realm of signal processing where the domain D of stream elements is continuous (say $D = \mathbb{R}$). In this case, it is possible to talk about low-pass filters which filter out some frequencies in a signal (which is thought to be a summation of waves with different frequencies). For a treatment of time-invariant filters within a quantitative-semantics framework see [36].

Considering the formulation of (DISTRIBUTION) it is a small step towards more specific axioms (FACTORING-N) that, for each $n \in \mathbb{N}$ with $n \geq 1$, capture exactly the n-window stream queries.

---

[8] An example use for this notion is given in the data mining book of Leskovec and colleagues [29].

(FACTORING-N) $\forall s \in D^*$: $Q(s) \in D^*$ and

1. if $|s| < n$, then $Q(s) = \epsilon$ and
2. if $|s| = n$, then for all $s' \in D^\infty$ with $|s'| \geq 1$: $Q(s \circ s') = Q(s) \circ Q((s \circ s')^{\geq 2})$.

Axiom (DISTRIBUTION) does not exactly correspond to (FACTORING-1) due to the special case of the empty stream. But the axiom (DISTRIBUTION-1) defined below does.

Now it is possible to prove results corresponding to propositions derived for Axiom (DISTRIBUTION), namely a proposition on the factorisation of the query result and the representability by a window-induced function.

Proposition 5.13 *For any $n \in \mathbb{N}$ with $n \geq 1$, a stream query $Q : D^\infty \longrightarrow D^\infty$ fulfilling (FACTORING-N) can be written for $|s| \geq n$ as*

$$Q(s) = \bigcirc_{j=1}^{|s|-n+1} Q(s^{[j,n+j-1]})$$

For $|s| < n$, $Q(s) = \epsilon$.

Proof See p. 127. □

Proposition 5.14 *For any $n \in \mathbb{N}$ with $n \geq 1$, a stream query $Q : D^\infty \longrightarrow D^\infty$ fulfils (FACTORING-N) iff it is induced by a normal n-window K.*

Proof See p. 127. □

Intuitively, the class of WAC stream queries is a proper class of AC stream queries because the former consider only fixed size finite portions of the input stream whereas for AC stream queries the whole past of an input stream is allowed to be incorporated for the production of the output stream. All WAC queries are bounded-memory processable in a very strict way. I discuss the point of bounded-memory processing in Sect. 5.6 in more detail and illustrate it here with the following example.

Example 5.15 A simple example for an AC query that is not a WAC query is the parity query *Parity* : $\{0, 1\}^\infty \longrightarrow \{0, 1\}^\infty$ which is defined as $Repeat(K_{par})$ where K_{par} is the parity window function $K : \{0, 1\}^* \longrightarrow \{0, 1\}$ given by

$$K_{par}(s) = \begin{cases} 1 & \text{if number of 1s in s is odd} \\ 0 & \text{else} \end{cases}$$

The window function K_{par} is not very complex, indeed one can show that K_{par} is bounded-memory computable. More concretely, it is easy to find a finite automaton with two states that accepts exactly those words with an odd number of 1s and rejects the others. In other words: parity is incrementally maintainable (to use a word from dynamic complexity [33]). But finite windows are "stateless", they cannot memorise the actual parity seen so far. Formally, it is easy to show that any finite-window function is AC^0 computable: For any word length m construct a circuit with m inputs where only the first n of them are actually used: one encodes all the 2^n values of the n-window K in a Boolean circuit BC_m, the rest of the m word is ignored. All BC_m

have the same size and depth and hence a finite window function is in AC^0. On the other hand, a classical result of Furst, Saxe, and Sipser [19] states that *Parity* is not in AC^0.

The axioms (FACTORING-N) are not exactly direct generalisations of the distribution axiom (DISTRIBUTION) as the $\circ$-factors do not factor disjoint parts of the input stream. A more intuitive set of axioms generalising (DISTRIBUTION) are the following axioms (DISTRIBUTION-N):

(DISTRIBUTION-N) For all $s \in D^*$: $Q(s) \in D^*$ and

1. if $|s| < n$, then $Q(s) = \epsilon$ and
2. if $|s| = n$, then for all $s' \in D^\infty$ with $|s'| \geq 1$: $Q(s \circ s') = Q(s) \circ Q(s')$.

Operators fulfilling (DISTRIBUTION-N) give the following factorisation representation of the output stream:

Proposition 5.16 *For any $n \in \mathbb{N}$ with $n \geq 1$, a stream query $Q : D^\infty \longrightarrow D^\infty$ fulfilling (DISTRIBUTION-N) can be written for $|s| \geq n$ as*

$$Q(s) = \bigcirc_{j=1}^{|s|} Q(s^{[nj-n+1,nj]})$$

Proof As above by induction. □

All stream queries fulfilling (DISTRIBUTION-N) can be characterised by *tumbling n-window windows* which are defined as functions $K : D^* \longrightarrow D^*$ such that

$$K(w) = \begin{cases} \epsilon & \text{if } |w| < n \\ \epsilon & \text{if } |w| \text{ is not a multiple of } n \\ K(v) & \text{if } |w| \text{ is a multiple with } w = uv \text{ and } |v| = n \end{cases}$$

Again note that one could have followed another approach by pushing the slide parameter into the definition of a new Repeat operator.

According to this definition, stream queries induced by tumbling windows consider only the last n elements in the stream and wait n elements before the content is updated. Using the words of the streaming community: the window has width n and slide n. According to this definition, in between times i and $i + n$ actually nothing is outputted to the output stream. This is one possible semantics for the slide action. It corresponds, in the speak of the relational data stream query language CQL [5], to the application of the IStream-operator to a sliding window operator. Another possibility would have been to output in between i and $i + n$ the contents of the last n-multiple i.

The following simple characterisation can be derived:

Proposition 5.17 *For any $n \in \mathbb{N}$ with $n \geq 1$, a stream query $Q : D^\infty \longrightarrow D^\infty$ fulfils (DISTRIBUTION-N) and the condition that $Q(s) = \epsilon$ for all $s \in D^*$ with $|s| < n$ iff it is induced by a tumbling n-window.*

Proof Follows with Proposition 5.16. □

Whereas (DISTRIBUTION-N) is a strengthening of (FACTORING-N) I consider now a generalisation of factoring which characterises stream queries induced by windows with a width n and slide factor m. I call such windows (n,m)-windows and define them as follows:

$$K(w) = \begin{cases} \epsilon & \text{if } |w| < max\{m,n\} \\ \epsilon & \text{if } |w| \text{ is not a multiple of } m \\ K(v) & \text{if } |w| \text{ is a multiple of m with } w = uv \text{ and } |v| = n \end{cases}$$

The corresponding axioms are

(GENERAL-FACTORING-N-M) $\forall s \in D^*$: $Q(s) \in D^*$ and

1. if $|s| < max\{n,m\}$, then $Q(s) = \epsilon$ and
2. if $|s| = n$, for all $s' \in D^\infty$ with $|s'| \geq 1$: $Q(s \circ s') = Q(s) \circ Q((s \circ s')^{\geq m})$.

5.5 Considering Time in the Word Model

A refined notion of abstract computability considered by Gurevich and colleagues [22] is that of *synchronous abstract computability, SAC* for short, which adds to the condition of abstract computability the condition that window K maps the empty stream to the empty stream and all non-empty streams to a finite stream of length 1.

Definition 5.18 A query Q is *synchronous abstract computable* (SAC) iff there is a window K with $K(\epsilon) = \epsilon$ and $|K(s)| = 1$ for all $s \neq \epsilon$ such that $Q(s) = Repeat(K)(s)$.

The authors give an axiomatic characterisation of this notion, this time using a property they call "non-predicting". The original definition (Def. 17) [22] says that Q is *non-predicting* iff for all streams s, s' and all $t \in \mathbb{N} \setminus \{0\}$ such that $s^{\leq t} = s'^{\leq t}$ one has $Q(s)^{=t} = Q(s')^{=t}$, which in natural speak says that the outcomes of the query at some time t depends only on the prefix of the streams up to t. Proposition 18 in [22] states that SAC queries are exactly the non-predicting queries.

There are some clarifications needed here, first regarding the reading of the definition, second regarding the chosen terminology, and third regarding the general conception of time in this model.

Regarding the reading of the definition it has to be stated explicitly that if s and s' have at least length of t and are the same up to t, then the outcomes $Q(s)$ and $Q(s')$ are also defined up to t and are the same at position t. So in particular a non-predicting operator maps infinite streams to infinite streams, i.e., being non-predicting entails (I2I). Considering their Proposition 18, the authors also assume that being non-predicting entails that finite streams are mapped to finite streams, i.e., that being non-predicting entails (F2F), because SAC queries map finite streams to finite streams. As I cannot find a plausible reading of non-predictability that entails this fact I rephrase their Proposition 18 by mentioning (F2F) explicitly.

Proposition 5.19 (Adaptation of [22]) *SAC queries are exactly the non-predicting queries that fulfill (F2F).*

Regarding the chosen terminology it should be noted that "non-predicting" is too weak a notion to capture the property stated under this term. The reason is that also a stream query Q_a that maps every non-empty stream (finite and infinite) to the constant stream (a) and the empty stream to itself does not produce streams by looking into the future—and hence Q_a should be called non-predicting in an intuitive sense. But surely Q_a is not SAC as it maps infinite streams to a finite stream.

So, intuitively it should be allowed to have the outcome up to some time point t being determined by elements later than the arrival of a stream element but not earlier. This leads to my notion of no-dilation computability: a query is Q is *no-dilation computable, for short: NDAC*, iff it is AC with a window K such that $K(\epsilon) = \epsilon$ and $K(s) \leq 1$ for all $s \in D^{\infty}$ with $s \neq \epsilon$. Choosing $K(\epsilon) = \epsilon$ and $K(s) = a$ for $s \neq \epsilon$ shows that query Q_a introduced above is an NDAC query.

The property corresponding to NDAC is a weakening of non-predictability (and a strengthening of FP^{∞}):

($\mathrm{FP}^{\infty}_{ND}$) For all $s \in D^{\infty}$ and all $u \in D^*$: if $Q(s) \in uD^{\infty}$, then there is a $w \in D^*$ such that $s \in wD^{\infty} \subseteq Q^{-1}(uD^{\infty})$ **and** $|w| \leq |u|$.

The following simple characterisation can be derived.

Proposition 5.20 *NDAC queries are exactly the queries fulfilling (FP^{∞}_{ND}).*

Proof Follows with the window construction from the proof of Theorem 5.4. □

Regarding the chosen time model it is clear that in the word model of Gurevich and colleagues [22] time is directly associated with the order in which the elements in a stream arrive. So time is modelled by the natural numbers with its natural ordering standing for the arrival ordering. In the synchronicity notion underlying the notion of SAC queries this time model is applied both to the input and the output stream. This synchronicity notion presupposes that the stream positions are considered as "absolute" time points: the second position in the output stream comes really later than the first position in the input stream, and the same positions in output and input stream denote the same time points. I am not going to discuss the plausibility of this notion of synchronicity but rather note that in the beginning of this chapter I defined a different notion of synchronicity based on an additional time dimension (for temporal) streams: the synchronicity aspect concerns the alignment of the arrival time with the application time and not the input-output synchronicity.

Having in the definition of temporal streams an additional time dimension is important when one is going to consider non-discrete time models where there may not be a smallest possible time interval in between to arriving elements. We refer the reader to [35] for further subtleties when considering continuous flows of time in stream processing.

I summarize all of the representation results achieved so far in Table 5.1

Stream query Q	**Axioms**
AC of form	
$Q: D^{\infty} \longrightarrow D^{\infty}$	(F2F) & (FP^{∞})
$Q: D^{\omega} \longrightarrow D^{\omega}$	(FP^{ω})
normal n-window	(FACTORING-N)
tumbling n-window	(DISTRIBUTION-N)
SAC	non-predicting & (F2F)
NDAC	($\mathrm{FP}^{\infty}_{ND}$)

Table 5.1 Representation Results

5.6 Bounded-Memory Queries

As mentioned before, the notion of abstract computability is very general, even so as to contain also queries that are not computable by a Turing machine according to the notion of TTE computability [42]. Hence, the authors of [22] consider the refined notion of *abstract computability modulo a class* C meaning that the window K inducing an abstract computable query has to be in C. In most cases C stands for a family of functions of some complexity class. In the paper, the authors consider variants of C based on computations by a machine model called *stream abstract state machine (sASM)*. In particular, they show that every AC query induced by a length-bounded window (in particular: each SAC query) is computable by an sASM [22, Corollary 23].

A particularly interesting class from the perspective of efficient computation are bounded-memory sASM because these implement the idea of incrementally maintainable windows requiring only a constant amount of memory. (For a more general notion of incremental maintainable queries see [33].) Of course these machines are quite restrictive as they, e.g., do not allow to compute the INTERSECT problem of checking whether prior to some given time point t there were identical elements in two given streams [22, Proposition 26]. A slightly more general version of bounded-memory sAMS are o(n)-bitstring sAMS which store, on every stream and every step, only o(n) bitstrings. But even these cannot compute INTERSECT [22, Proposition 28].

I give a rough sketch of the ASM and the sASM machine model, referring the reader to [22] for details. An ASM and sASM operates on first-order sorted FOL structures with a static part and a dynamic part. The dynamic part consists of functions which may change by transitions in an update process. Updates are the basic transitions. Based on these, simple programs are defined as finite sequences of rules: the basic rules are updates $f(t_1,\ldots,t_n) := t_0$. Then, inductively, a rule is built by applying to basic rules a parallel execution construct or by applying on given rules r_1, r_2 an if-then-else construct if Q then r_1 else r_2, where the if-condition is given by a quantifier free formula Q on the signature of the structure and where the post-conditions are r_1, r_2.

The machine model of ASM and sASM do not seem to be widely used in the sensor network stream community, where there is no complicated FOL structure to manipulate, nor the database (stream) community where there is a similar but slimmer model termed *relational transducer* [1]. The authors of [22] already hint to a connection between the sAsm model and the model of Finite Cursor Machines (FCM) [21], which are more expressive than sAsms as they may have different cursors which are controlled on the client side and hence are not data-driven. In fact, in Proposition 29 [22] the authors describe a problem that is computable by a FCM but not by a bounded-memory sAsm.

Though the machine-oriented approach for the characterisation of bounded-memory processing is quite universal and fits into the general approach for characterising computational classes, I add here a simple, straight-forward characterisation following the idea of primitive recursion over words [24, 7]: starting from basic functions on finite words, the user is allowed to built further functions by applying composition and simple forms of recursion. In order to guarantee bounded memory processing, all the construction rules are built with specific window operators, namely $last_n(\cdot)$, which outputs the n-suffix. This construction enables user to built bounded-memory window functions K in a pipeline strategy. The main extension to the approach of [22] is adding recursion, which leads to a fine-grained definition of kernels K.

In order to enable a pipeline-based construction I extend further the approach of [22] by considering multiple streams explicitly as possible arguments for functions with an arbitrary number of arguments. Still, all functions will output a single finite or infinite word—though the approach sketched below can easily be adapted to work for multi-output streams. All of the machinery of Gurevich's framework is easily translated to this multi-argument setting. So, for example (FP^∞) now reads as follows:

(FP^∞) For all $s_1, \dots s_n \in D^\infty$, and all $u \in D^*$: if $Q(s_1, \dots, s_n) \in uD^\infty$, then there are $w_1, \dots, w_n \in D^*$ such that $s_i \in w_i D^\infty$ for all $i \in [n]$ and it holds that $w_1 D^\infty \times \dots \times w_n D^\infty \subseteq Q^{-1}(uD^\infty)$.

Monotonicity of a function $Q : (D^\infty)^n \longrightarrow D^\infty$ now reads as: for all $(s_1, \dots, s_n)$ and $(s'_1, \dots, s'_n)$ with $s_i \sqsubseteq s'_i$ for all $i \in [n]$: $Q(s_1, \dots, s_n) \sqsubseteq Q(s'_1, \dots, s'_n)$.

In a second step I further adapt the set of rules with a co-recursive rule, in order to describe directly bounded-memory queries $Q = Repeat(K)$—instead of only the underlying windows K.

I define three types of classes in parallel: classes Accu^n which are intended to model accumulator functions $f : (D^*)^n \longrightarrow D^*$; classes $\text{Mbinc}^{(n;m)}$ that model incrementally maintainable functions with bounded memory, i.e., bounded-memory window functions with bounded output, and classes $\text{Mmbinc}^{(n;m)}$ of incrementally maintainable, bounded-memory , and monotonic functions that lead to the definition of monotonic functions on infinite streams. The main idea, similar to that of [7], is to partition the argument functions in two classes, normal and safe arguments. In [7] the normal variables are the ones on which the recursion step happens and which have to be controlled, whereas the safe ones are those in which the growth of the term is not restricted. Here I control the growth (the length) of the words explicitly and use rather the distinction between input and output arguments: the input arguments are those where the input may be either a finite or an infinite word. The output variables are the ones in which the accumulation happens. In a function term $f(x_1, \ldots, x_n; y_1, \ldots, y_m)$ the input arguments are the ones before the semicolon ";", here: $x_1, \ldots, x_n$, and the output arguments are the ones after the ";", here $y_1, \ldots, y_n$.

Using the notation of [24] for my purposes, a function f with n input and m output arguments is denoted $f^{(n;m)}$. Classes $\text{MBINC}^{(n;m)}$ and $\text{MMBINC}^{(n;m)}$ consist of functions of the form $f^{(n;m)}$. The class $\text{MMBINC} = \bigcup_{n \in \mathbb{N}} \text{MMBINC}^{(n;)}$ contains all functions without output variables and is the class of functions which describe the prefix restrictions $Q_{\restriction D^*}$ of stream queries $Q : D^\infty \longrightarrow D^\infty$ that are computable by a bounded-memory sAsm.

Definition 5.21 The set of *bounded n-ary accumulator word functions*, for short Accu^n, the set of $(n+m)$-ary *bounded-memory incremental functions* with n input and m output arguments, for short $\text{MBINC}^{(n;m)}$, and the set of *monotonic, bounded-memory , incremental* $(n+m)$*-ary functions* with n input and m output arguments, for short $\text{MMBINC}^{(n;m)}$, are defined according to the following rules:

1. $w \in \text{Accu}^0$ for any word $w \in D^*$ ("Constants")
2. $last_k(\cdot) \in \text{Accu}^1$ for any $k \in \mathbb{N}$ ("Suffixes")
3. $S_k^a(w) = last_k(w) \circ a \in \text{Accu}^1$ for any $a \in D$ ("Successors")
4. $P_k(w) = last_{k-1}(w) \in \text{Accu}^1$ ("Predecessors")
5. $cond_{k,l}(w, v, x) = \begin{cases} last_k(v) & \text{if } last_1(w) = 0 \\ last_l(x) & \text{else} \end{cases} \in \text{Accu}^3$ ("Conditional")
6. $\Pi_k^j(w_1, \ldots, w_n) = last_k(w_j) \in \text{Accu}^n$ for any $k \in \mathbb{N}$ and $j \in [n]$, $n \neq 0$. ("Projections")
7. $shl(\cdot)^{(1;0)} \in \text{MMBINC}$ with $shl(aw;) = w$ and $shl(\epsilon;) = \epsilon$. ("Left shift")
8. Conditions for Composition ("Composition")

   a. If $f \in \text{Accu}^n$ and, for all $i \in [n]$, $g_i \in \text{Accu}^m$, then also $f(g_1, \ldots, g_n) \in \text{Accu}^m$; and:

   b. If $g^{(m;n)} \in \text{MMBINC}^{(m;n)}$ and, for all $i \in [m]$, $g_i \in \text{Accu}^l$ and $h_j^{(k;l)} \in \text{MMBINC}^{(k;m)}$ for $j \in [n]$, then $f^{(k;l)} \in \text{MMBINC}^{(k;l)}$ where using $\mathbf{w} = w_1, \ldots, w_k$, $\mathbf{v} = v_1, \ldots, v_l$

   $$f^{(k;l)}(\mathbf{w}; \mathbf{v}) = g^{(m;n)}(h_1(\mathbf{w}; \mathbf{v}), \ldots, h_m(\mathbf{w}; \mathbf{v}); g_1(\mathbf{v}), \ldots, g_n(\mathbf{v}))$$

   c. If $g^{(m;n)} \in \text{MBINC}^{(m;n)}$ and, for all $i \in [m]$, $g_i \in \text{Accu}^l$ and $h_j^{(k;l)} \in \text{MMBINC}^{(k;m)}$ for $j \in [n]$, then $f^{(k;l)} \in \text{MBINC}^{(k;l)}$ where using $\mathbf{w} = w_1, \ldots, w_k$, $\mathbf{v} = v_1, \ldots, v_l$

   $$f^{(k;l)}(\mathbf{w}; \mathbf{v}) = g^{(m;n)}(h_1(\mathbf{w}; \mathbf{v}), \ldots, h_m(\mathbf{w}; \mathbf{v}); g_1(\mathbf{v}), \ldots, g_n(\mathbf{v}))$$

9. If $g : (D^*)^n \longrightarrow D^* \in \text{Accu}$ and $h : (D^*)^{n+3} \longrightarrow D^* \in \text{Accu}$ then also $f : (D^*)^{n+1} \longrightarrow D^* \in \text{Accu}$, where:

$$f(\epsilon, v_1, \ldots, v_n) = g(v_1, \ldots, v_n)$$
$$f(wa, v_1, \ldots, v_n) = h(w, a, v_1, \ldots v_n, f(w, v_1, \ldots, v_n))$$

("Accu-Recursion")

10. If $g_i : (D^*)^{n+m} \longrightarrow D^* \in$ ACCU for $i \in [m]$, $g_0 \in$ ACCU then $k = k^{(n;m)} \in$ MBINC$^{(n;m)}$, where k is defined using the above abbreviations as follows:

$$k(\epsilon, \ldots, \epsilon; \mathbf{v}) = g_0(\mathbf{v})$$
$$k(\mathbf{w}; \mathbf{v}) = k(shl(\mathbf{w}); g_1(\mathbf{v}, \mathbf{w}^{=1}), \ldots, g_m(\mathbf{v}, \mathbf{w}^{=1}))$$

("Window-Recursion")

11. If $g_i : (D^*)^{n+m} \longrightarrow D^* \in$ ACCU for $i \in [m]$, $g_0 \in$ ACCU, then $f = f^{(n;m)} \in$ MMBINC$^{(n;m)}$, where f is defined using the above abbreviations as follows:

$$f(\epsilon, \ldots, \epsilon; out, \mathbf{v}) = out$$
$$f(\mathbf{w}; out, \mathbf{v}) = f(shl(\mathbf{w}); out \circ g_1(\mathbf{v}, \mathbf{w}^{=1}), g_1(\mathbf{v}, \mathbf{w}^{=1}), \ldots, g_m(\mathbf{v}, \mathbf{w}^{=1}))$$

("Repeat-Recursion")

Let MMBINC $= \bigcup_{n \in \mathbb{N}}$ MMBINC$^{(n;)}$.

Within the definition above three types of recursions occur: the first is a primitive recursion over accumulators. The second, called window-recursion, is a specific form of *tail recursion* which means that the recursively defined function is the last application in the recursive call. As the name indicates, this recursion rule is intended to model the window functions. The last recursion rule (again in tail form) is intended to mimic the *Repeat* functional.

In the first recursion, the word is consumed from the end: this is possible, as the accumulators are built from left to right during the streaming process. Note, that the length of outputs produced by the accu-recursion rule and the window-recursion rule are length-bounded.

The window-recursion rule and the repeat-recursion rule implement a specific form of tail recursion consuming the input words from the beginning with the left-shift function *shl*(). This is required as the input streams are potentially infinite. Additionally, these two rules implement a modified form of simultaneous recursion, where all input words are consumed in parallel. The temporal model behind this form of recursion is the following: At every time point one has exactly n elements to consume, exactly one for each of the n input streams. These are thought to appear at the same time. To model also the case where no element arrives in some input stream, a specific symbol $\perp$ can be added to the system. Giving the engine a finite word as input means that the engine gets noticed about the end of the word (when he has read the word). Note then, that there is a difference between the finite word abc and the infinite word $abc(\perp)^\omega$.

Repeat recursion is illustrated with the following simple example.

Example 5.22 Consider the window function K_{par} that, for a word w, outputs its parity. The monotonic function $Par(w) = Repeat(K_{par})(w) = \bigcirc_{j=0}^{|w|} K_{par}(w^{\leq j})$

can be modelled as follows. The auxiliary xor function $\oplus$ can be defined with cond because with cond one can define the functionally complete set of junctions $\{\neg, \wedge\}$ with $\neg x := cond_{1,1}(x, 1, 0)$ and $x \wedge y = cond_{1,1}(x, 0, y)$. Using repeat recursion (item 11 in Definition 5.21) gives the desired function.

$$\begin{aligned} f(\epsilon; out, v) &= out \\ f(w; out, v) &= f(shl(w); out \circ v \oplus w^{=1}, v \oplus w^{=1}) \\ Par(w) &= f(w; \epsilon, 0) \end{aligned}$$

For example, the input word $w = 101$ is consumed as follows:

$$\begin{aligned} Par(101) &= f(101; \epsilon, 0) = f(shl(101); \epsilon \circ 0 \oplus 101^{=1}, 0 \oplus 101^{=1}) \\ &= f(01; \epsilon \circ 0 \oplus 1, 0 \oplus 1) = f(01; 1, 1) \\ &= f(1; 1 \circ 1 \oplus 0, 1 \oplus 0) = f(1; 1 \circ 1, 1) \\ &= f(\epsilon; 1 \circ 1 \oplus 1, 1 \oplus 1) = f(\epsilon; 1 \circ 1 \circ 0, 0) = 110 \end{aligned}$$

The output of the repeat-recursion grows linearly: the whole history is outputted with the help of the concatenation function. Note, by the way, that the concatenation functions appears only in the repeat-recursion rule and also in a restricted form in the successor functions, but there is no concatenation function defined in one of the three classes. The recursion anchor has the special form that it directly outputs the first output position of the recursively defined function f. Because of this, it follows that all functions in MMBINC are monotonic in their input arguments. I state this as a proposition:

Proposition 5.23 *All functions in MMBINC are monotonic*

Proof I introduce the notion of a function $f(\mathbf{x}; \mathbf{y})$ being monotonic w.r.t. its arguments $\mathbf{x}$: This is the case if for every $\mathbf{y}$ the function $f_{\mathbf{y}}(\mathbf{x}) = f(\mathbf{x}, \mathbf{y})$ is monotonic.

The functions in MMBINC are either the left shift function (which is monotonic) or a function constructed with the application of composition, which preserves monotonicity, or by repeat-recursion, which, due to the concatenation in the output position, also guarantees monotonicity. □

The functions in MMBINC map (vectors of) finite words to finite words. Because of the monotonicity, it is possible to define for each $f \in$ MMBINC an extension $\tilde{f}$ which maps (vectors) of finite or infinite words to finite or infinite words. If $f^{(n;)} : (D^*)^n \longrightarrow D^*$, then $\tilde{f} : (D^\infty)^n \longrightarrow D^\infty$ is defined as follows: if all $s_i \in D^*$, then $\tilde{f}(s_1, \ldots, s_n) = f(s_1, \ldots, s_n)$. Otherwise, $\tilde{f}(s_1, \ldots, s_n) = sup_{i \in \mathbb{N}} f(s_1^{\leq i}, \ldots, s_n^{\leq i})$ where $sup_{i \in \mathbb{N}} f(s_1^{\leq i}, \ldots, s_n^{\leq i})$ is the unique stream $s \in D^\infty$ such that $f(s_1^{\leq i}, \ldots, s_n^{\leq i}) \sqsubseteq s$ for all i. I denote by BMSQ those functions Q that can be presented as $Q = \tilde{f}$ for some $f \in$ MMBINC and call them bounded-memory stream queries.

Theorem 5.24 *A function Q with one argument belongs to BMSQ iff it is a stream query computable by a bounded-memory sASM.*

Proof See p. 128. □

Note that in a similar way one can model $o(n)$ bitstring bounded sAsm: instead of using constant size windows $last_k(c)$ in the definition of accumulator functions, one uses dynamic windows $last_{f(\cdot)}(\cdot)$, where, for a sublinear function $f \in o(n)$, $last_{f(|w|)}(w)$ denotes the $f(|w|)$ suffix of w.

5.7 Related Work

The work presented here is based on the foundation of stream processing according to [22] which considers streams as finite or infinite words. The research on streams from the word perspective—sometimes referred to as stringology—is quite mature, and the literature on infinite words, language characterisations, and associated machine models abounds. The focus in this chapter is on representational aspects for functions from words to words. For all other interesting topics and relevant research papers on infinite words we refer the reader to [42] and [34].

The representation theorems in this chapter are based on the *Repeat* functional and a window function. An alternative representation by trees is given in [23]: An (infinite) input word is read as sequence of instructions to follow the tree, 0 for left and 1 for right. The leaves of the tree contain the elements to be outputted. The authors give a characterisation for the interesting case where the range of the stream query is a set of infinite words. In this case they have to use non-well-founded trees. Note, that in this type of representation the construction principle becomes relevant. Instead of a simple instantiation with a parameter value, one has to apply an algorithm in order to build the structure (here: the function).

In [22] and in this chapter, the underlying alphabet for streams is not necessarily finite. This is similar to the situation in research on data words [39, 9], where the elements of the stream have next to an element from a finite alphabet also an element from an infinite alphabet.

The approach of this chapter (and of the whole monograph) is axiom based. An example of an axiom-based approach for stream processing is given in [35], but in [35] the emphasis is on temporal streams over continuous flows of time (such as the reals).

Until now streams are considered in an abstract way, with stream elements over an arbitrary domain D. In higher-level stream processing, where the domain elements have a certain semantics according to specifications in a knowledge base or an ontology, further aspects related to semantics become relevant. Three of them will be dealt with in the following section on the STARQL framework: correct window semantics w.r.t. the given certain-answer semantics, reasoning aspects w.r.t. streamed OBDA, and the equivalence of stream queries of different stream languages. Regarding reasoning aspects I refer the reader also to [25] and regarding the equivalence of stream queries the reader may find [6] useful.

Aspects of performant—in particular bounded-memory—processing on streams are touched in this chapter with the construction of a class of functions capturing exactly those queries computable by an sAsm. This characterisation is in the tradition of implicit complexity as developed in the PhD thesis of Bellantoni [8] which is based on work of Leivant [28]. (See also the summary of the thesis in [7] where the main result is the characterisation of polynomial time functions by some form of primitive recursion). The main idea of distinguishing between two sorts of variables in my approach comes from [8], the use of constant or $o(n)$ sized windows to control the primitive recursion is similar to the approach of [30] used for the "bounded recursion" rule.

The consideration of bounded-memory computations in [4] is couched in the terminology of data-stream management systems. The authors of [4] consider FOL or rather: (non-recursive) SQL as the language to represent windows. The main result is a syntactical criterion for deciding whether a given FOL formula represents a bounded-memory query. Similar results in the tradition of Büchi's result on the equivalence of finite-automata recognisability with definability in second-order logic over the sequential calculus can be shown for streams in the word perspective [16, 2, 3, 18, 12].

An aspect related to bounded-memory processing is that of incremental maintainability as discussed in the area called dynamic complexity [33, 43]. Here the main concern is to break down a query on a static data set into a stream query using simple update operators with small space.

The function-oriented consideration of stream queries along the line of this chapter and [22] lends itself to a pipeline-style functional programming language on streams. And indeed, there are some a few examples, such as [10], that show the practical realisability of such a programming language.

The type of recursion I have used in order to handle infinite streams, namely the rules of window-revision and repeat-revision, uses the consumption of words from the beginning. This is similar to the co-algebraic approach for defining streams and stream functions [37, 38, 14].

5.8 Summary

With the general (infinite) word-based framework I considered a sufficiently general, yet simple model of streams which allows for specifying properties of stream queries and characterising them. Though the achieved results have a foundational character, they are useful for applications relying, e.g., on the agent paradigm where stream processing plays an important role. In setting up the stream (reasoning) architecture of an agent, the engineer can rely on these results in order to ensure a specific input-output behaviour when using particular classes of stream queries.

The axiomatic characterisations given in this chapter are on a basic phenomenological level—phenomenological, because only observations regarding the input-output behaviour are taken into account, and basic, because no further properties regarding the structure of the data stream elements are presupposed. So, the axiomatic characterisations of this chapter lay the ground and are the starting point for a more elaborated characterisation of rational agents where also the properties of various higher-order streams of states, beliefs, and goals are taken into account. Such an elaborated characterisation of rational agents requires inventing axioms referring to the specific observable properties associated with a stream of a given type.

For example, considering the stream of epistemic states $\Phi_1, \Phi_2, \ldots$ of an agent, an associated observable property is the set of beliefs $Bel(\Phi_i)$ an agent is obliged to believe in its current state Φ_i. The beliefs can be expressed in some logic which comes with an entailment relation $\models$. Using the entailment relation, the idea of a rational change of beliefs of the agent under new information can be made precise.

In order to illustrate the kind of intended elaboration, I mention here the success axiom which expresses that the agent "trusts" in the information it receives: if the agent receives new information α, then the current state Φ_i should develop into state Φ_{i+1} such that $Bel(\Phi_{i+1}) \models \alpha$. The constraining effects that this axiom has on the belief-state change may appear simple but, at least when the new information is not consistent with the current beliefs, it is not clear how the change has to be carried out. Axioms such as the success axiom are one of the main objects of study in the field of belief revision (compare the introductory chapter and Chap. 7.) But what is still missing in current research is the combination of belief-revision axioms with axioms expressing basic stream properties—such as those discussed in this chapter.

Though the subfield of belief-revision that deals with iterated applications of a revision operator on a stream of new information seems to be the ideal framework for the mentioned combination of revision axioms and stream axioms, there has been not much progress along this line since the landmarking paper on iterated belief revision by Darwiche and Pearl [11].

The intended elaboration aimed at is comparable in its generality with the idea of universal AI as outlined in [26]. But whereas [26] is based mainly on considerations from algorithmic probability, the elaboration I have in mind is in the symbolic tradition of logics and belief revision.

References

1. Abiteboul, S., Vianu, V., Fordham, B., Yesha, Y.: Relational transducers for electronic commerce. Journal of Computer and System Sciences **61**(2), 236 – 269 (2000). DOI http://dx.doi.org/10.1006/jcss.2000.1708
2. Alur, R., Cerný, P.: Expressiveness of streaming string transducers. In: K. Lodaya, M. Mahajan (eds.) Proceedings of the Annual Conference on Foundations of Software Technology and Theoretical Computer Science (FSTTCS-10), *LIPIcs*, vol. 8, pp. 1–12. Schloss Dagstuhl - Leibniz-Zentrum für Informatik (2010). DOI 10.4230/LIPIcs.FSTTCS.2010.1
3. Alur, R., Filiot, E., Trivedi, A.: Regular transformations of infinite strings. In: Proceedings of the 27th Annual IEEE/ACM Symposium on Logic in Computer Science (LICS-12), LICS '12, pp. 65–74. IEEE Computer Society, Washington, DC, USA (2012). DOI 10.1109/LICS.2012.18
4. Arasu, A., Babcock, B., Babu, S., McAlister, J., Widom, J.: Characterizing memory requirements for queries over continuous data streams. ACM Transactions on Database Systems **29**(1), 162–194 (2004). DOI 10.1145/974750.974756
5. Arasu, A., Babu, S., Widom, J.: The CQL continuous query language: Semantic foundations and query execution. The VLDB Journal **15**, 121–142 (2006)

6. Beck, H., Dao-Tran, M., Eiter, T.: Equivalent stream reasoning programs. In: S. Kambhampati (ed.) Proceedings of the 25th International Joint Conference on Artificial Intelligence (IJCAI-16), pp. 929–935. IJCAI/AAAI Press (2016)
7. Bellantoni, S., Cook, S.: A new recursion-theoretic characterization of the polytime functions. Computational Complexity **2**(2), 97–110 (1992). DOI 10.1007/BF01201998
8. Bellantoni, S.J.: Predicative recursion and computation complexity. Ph.D. thesis, Graduate Department of Computer Science, University of Toronto (1992)
9. Benedikt, M., Ley, C., Puppis, G.: Automata vs. logics on data words. In: A. Dawar, H. Veith (eds.) Computer Science Logic, *LNCS*, vol. 6247, pp. 110–124. Springer Berlin Heidelberg (2010). DOI 10.1007/978-3-642-15205-4_12
10. Cowley, A., Taylor, C.J.: Stream-oriented robotics programming: The design of roshask. In: Proceedings of the 2011 IEEE/RSJ International Conference on Intelligent Robots and Systems (IROS-11), pp. 1048–1054 (2011). DOI 10.1109/IROS.2011.6095033
11. Darwiche, A., Pearl, J.: On the logic of iterated belief revision. Artificial Intelligence **89**, 1–29 (1997)
12. Dave, V., Narayanan Krishna, S., Trivedi, A.: FO-definable transformations of infinite strings. ArXiv e-prints (2016)
13. Della Valle, E., Ceri, S., van Harmelen, F., Fensel, D.: It's a streaming world! Reasoning upon rapidly changing information. IEEE Intelligent Systems **24**(6), 83 –89 (2009). DOI 10.1109/MIS.2009.125
14. Endrullis, J., Grabmayer, C., Hendriks, D., Isihara, A., Klop, J.W.: Productivity of stream definitions. Theoretical Computer Science **411**(4-5), 765–782 (2010). DOI 10.1016/j.tcs.2009.10.014
15. Endrullis, J., Hendriks, D., Klop, J.W.: Streams are forever. Bulletin of the EATCS **109**, 70–106 (2013)
16. Engelfriet, J., Hoogeboom, H.J.: Mso definable string transductions and two-way finite-state transducers. ACM Transationcs on Computational Logic **2**(2), 216–254 (2001). DOI 10.1145/371316.371512
17. Eschenbach, C., Özçep, Ö.L.: Ontology revision based on reinterpretation. Logic Journal of the IGPL **18**(4), 579–616 (2010). DOI doi:10.1093/jigpal/jzp039. First published online August 12, 2009
18. Filiot, E., Gauwin, O., Reynier, P.A., Servais, F.: From two-way to one-way finite state transducers. In: Proceedings of the 28th Annual ACM/IEEE Symposium on Logic in Computer Science (LICS-13), pp. 468–477. IEEE Computer Society, Washington, DC, USA (2013). DOI 10.1109/LICS.2013.53
19. Furst, M., Saxe, J.B., Sipser, M.: Parity, circuits, and the polynomial-time hierarchy. Theory of Computing Systems **17**(1), 13–27 (1984)
20. Gries, O., Möller, R., Nafissi, A., Rosenfeld, M., Sokolski, K., Wessel, M.: A probabilistic abduction engine for media interpretation based on ontologies. In: J. Alferes, P. Hitzler, T. Lukasiewicz (eds.) Proceedings of the 4th International Conference on Web Reasoning and Rule Systems (RR-10) (2010)
21. Grohe, M., Gurevich, Y., Leinders, D., Schweikardt, N., Tyszkiewicz, J., Van den Bussche, J.: Database query processing using finite cursor machines. Theory of Computing Systems **44**(4), 533–560 (2009). DOI 10.1007/s00224-008-9137-7
22. Gurevich, Y., Leinders, D., Van Den Bussche, J.: A theory of stream queries. In: Proceedings of the 11th International Conference on Database Programming Languages, DBPL'07, pp. 153–168. Springer-Verlag, Berlin, Heidelberg (2007)
23. Hancock, P., Pattinson, D., Ghani, N.: Representations of Stream Processors Using Nested Fixed Points. Logical Methods in Computer Science **5**(3:9), 1–17 (2009)
24. Handley, W.G., Wainer, S.S.: Complexity of primitive recursion. In: U. Berger, H. Schwichtenberg (eds.) Computational Logic: Proceedings of the NATO Advanced Study Institute on Computational Logic, held in Marktoberdorf, Germany, July 29–August 10, 1997, pp. 273–300. Springer Berlin Heidelberg, Berlin, Heidelberg (1999). DOI 10.1007/978-3-642-58622-4_8

25. Heintz, F., Kvarnström, J., Doherty, P.: Bridging the sense-reasoning gap: Dyknow - stream-based middleware for knowledge processing. Advanced Engineering Informatics **24**(1), 14–26 (2010). DOI 10.1016/j.aei.2009.08.007
26. Hutter, M.: Universal Artificial Intelligence: Sequential Decisions Based on Algorithmic Probability. Springer-Verlag, Berlin, Heidelberg (2010)
27. Krämer, J., Seeger, B.: Semantics and implementation of continuous sliding window queries over data streams. ACM Trans. Database Syst. **34**(1), 1–49 (2009). DOI 10.1145/1508857.1508861
28. Leivant, D.: A foundational delineation of poly-time. Information and Computation **110**(2), 391–420 (1994). DOI http://dx.doi.org/10.1006/inco.1994.1038
29. Leskovec, J., Rajamaran, A., Ullman, J.D.: Mining of Massive Datasets. Cambridge University Press (2014)
30. Lind, J., Meyer, A.R.: A characterization of log-space computable functions. SIGACT News **5**(3), 26–29 (1973). DOI 10.1145/1008293.1008295
31. Özçep, Ö.L.: Bounded-memory stream processing. In: Proceedings of the 41st German AI Conference (KI 2018), Berlin (2018). Accepted for publication
32. Özçep, Ö.L., Möller, R.: Towards foundations of agents reasoning on streams of percepts. In: Proceedings of the 31st International Florida Artificial Intelligence Research Society Conference (FLAIRS-18), pp. 80–85 (2018)
33. Patnaik, S., Immerman, N.: Dyn-fo: A parallel, dynamic complexity class. Journal of Computer and System Sciences **55**(2), 199–209 (1997)
34. Perrin, D., Pin, J.: Infinite Words: Automata, Semigroups, Logic and Games. Pure and Applied Mathematics. Elsevier Science (2004)
35. Rabinovich, A.M.: Automata over continuous time. Theoretical Computer Science **300**(1-3), 331–363 (2003)
36. Rodionova, A., Bartocci, E., Nickovic, D., Grosu, R.: Temporal logic as filtering. In: A. Abate, G.E. Fainekos (eds.) Proceedings of the 19th International Conference on Hybrid Systems: Computation and Control (HSCC-16), pp. 11–20. ACM (2016). DOI 10.1145/2883817.2883839
37. Rutten, J.: Elements of stream calculus (an extensive exercise in coinduction). Electronic Notes in Theoretical Computer Science **45**(0), 358–423 (2001). DOI http://dx.doi.org/10.1016/S1571-0661(04)80972-1
38. Rutten, J.J.M.M.: A coinductive calculus of streams. Mathematical Structures in Computer Science **15**(1), 93–147 (2005). DOI 10.1017/S0960129504004517
39. Segoufin, L.: Automata and logics for words and trees over an infinite alphabet. In: Z. Ésik (ed.) Proceedings of the 20th International Conference on Computer Science Logic (CSL-06), *LNCS*, vol. 4207, pp. 41–57. Springer (2006). DOI 10.1007/11874683_3
40. Tucker, P.A., Maier, D., Sheard, T., Fegaras, L.: Exploiting punctuation semantics in continuous data streams. IEEE Transactions on Knowledge and Data Engineering **15**(3), 555–568 (2003). DOI 10.1109/TKDE.2003.1198390
41. Valle, E.D., Schlobach, S., Krötzsch, M., Bozzon, A., Ceri, S., Horrocks, I.: Order matters! Harnessing a world of orderings for reasoning over massive data. Semantic Web **4**(2), 219–231 (2013)
42. Weihrauch, K.: Computable Analysis: An Introduction. Springer-Verlag New York, Inc., Secaucus, NJ, USA (2000)
43. Zeume, T., Schwentick, T.: Dynamic conjunctive queries. In: N. Schweikardt, V. Christophides, V. Leroy (eds.) Proceedings of the 17th International Conference on Database Theory (ICDT-14), pp. 38–49. OpenProceedings.org (2014). DOI 10.5441/002/icdt.2014.08

Appendix

Proof of Theorem 5.3 (mainly [22])

First assume that Q is AC. Then it clearly maps finite streams to finite streams. Moreover, it fulfils (FP$^\infty$): For $s \in D^\infty$ and $u \in D^*$ assume that $Q(s) \in uD^\infty$, so there is an $l \in \mathbb{N}$ such that $u \sqsubseteq \bigcirc_{j=0}^{l} Q(s^{\leq j})$, Consider $w := s^{\leq l}$. Then $s \in wD^\infty$ and for all $s' \in wD^\infty$ one has $Q(s') \sqsupseteq Q(w)$ and hence $Q(s') \in uD^\infty$.

Now assume that Q maps finite streams to finite streams and that Q is AC. Define a window K as follows

$$K(\epsilon) := \epsilon$$
$$K(sa) = Q(sa) -_{\sqsubseteq} Q(s) \text{ for } s \in D^*, a \in D$$

In order to show that K is well-defined it has to be shown that $Q(s) \sqsubseteq Q(sa)$. But we have $Q(s) \in D^\infty$, so due to (FP$^\infty$) there is $w \in D^\infty$ such that $s \in wD^\infty \subseteq Q^{-1}(Q(s)D^\infty)$. So as $sa \sqsupseteq s$ one has also $Q(sa) \in f(s)D^\infty$, in other words: $Q(s) \sqsubseteq Q(sa)$.

Now one has to show that $Repeat(K)$ indeed is the same as Q. This is shown by showing that both functions agree on all streams. This is by definition true for finite streams. So let $s \in D^\omega$ be an infinite stream. First, I show that any prefix of $Repeat(K)$ is a prefix of Q. Let and $v \sqsubseteq Repeat(Q)$ and assume that i is the smallest i such that $v \sqsubseteq Repeat(Q)(s^{\leq i})$. But $Repeat(Q)(s^{\leq i}) = Q(s^{\leq i}) \in Q(s^{\leq i})D^\infty$. Hence, as Q fulfils (FP$^\infty$) it follows that there is $w \in D^*$ such that $s^{\leq i} \in wD^\infty \subseteq Q^{-1}(Q(s^{\leq i})D^\infty)$. Hence $Q(s^{\leq i}) \sqsubseteq Q(s)$. Now assume that v is a prefix of $Q(s)$. This means that $f(s) \in vD^\infty$. So, again using (FP$^\infty$), it follows that there is a w such that $s \in wD^\infty \subseteq Q^{-1}(vD^\infty)$. That is $Q(wD^\infty) \subseteq vD^\infty$ hence $v \sqsubseteq Q(ws')$ for any $s' \in D^\infty$. In particular $Q(w) \in vD^\infty$, but $Repeat(K)(w) = Q(w)$, so v is also a prefix of $Repeat(K)(s)$ (because $w \sqsubseteq s$).

Proof of Theorem 5.4

The proof that AC functions $Q : D^\omega \longrightarrow D^\omega$ fulfill (FP$^\omega$) proceeds in the same way as for functions $Q : D^\infty \longrightarrow D^\infty$ in the proof of Theorem 5.3.

For the other direction assume that $Q : D^\omega \longrightarrow D^\omega$ fulfils (FP$^\omega$).

Let s be any infinite stream $s \in D^\omega$. Consider an ordering of all growing prefixes u_i of $Q(s)$ with $u_0 = \epsilon$ and $u_i = (Q(s))^{\leq i}$. According to (FP$^\omega$) there is, for each u_i, a $w \in D^*$ such that

$$s \in wD^\omega \subseteq Q^{-1}(u_i D^\omega)$$

One may assume that for each u_i one chooses its w as the smallest such word w w.r.t. the prefix order. We call the sequence of resulting words $(w_i^s)_{i \in \mathbb{N}}$. Because of the minimality it follows that the sequence is increasing w.r.t. the prefix order. It may be the case that $w_i^s = w_{i+1}^s$. So I consider the set of indexes $H \subseteq \mathbb{N}$ such that for all $j \in H$ it holds that $w_j \sqsubseteq w_{j+1}$ but $w_j \neq w_{j+1}$, for short if $w_j \sqsubset w_{j+1}$. Assume that H is given as a family $H = (i_k)_{k\in Y}$ where $Y = \mathbb{N}$ or $Y = \{1, \ldots, n\}$ for some natural number n. Now we define the following function K^s for all words $v \sqsubseteq s$. The word v can be of the following form: It is some word where there is a proper growth from w_j to w_{j+1}, i.e., it is a word of the form w_{i_k} or it is a word of the form w_j where no change happens, than it has one of the form $w_{i_k+1}, w_{i_k+2}, \ldots, w_{i_{k+1}-1}$. Or v is not represented as a w_j.

$$K^s(w_j^s) = \begin{cases} u_{i_{k+1}-1} & \text{if } j \in \{i_0, i_0+1, \ldots, i_{0+1}-1\} \text{ and } k \geq 1 \\ u_{i_{k+1}-1} -_{\sqsubseteq} K(w_{i_{k-1}}^s) & \text{if } j \in \{i_k, i_k+1, \ldots, i_{k+1}-1\} \text{ and } k \geq 1 \end{cases}$$

$$K^s(w_j^s u) = \epsilon \quad \text{if } u \in D^* \text{ and } w_j^s \sqsubset w_j^s u \sqsubset w_{j+1}$$

Then by definition we have that

$$Repeat(K^s)(s) = Q(s)$$

Of course this is still not the window K that is required because K^s still depends on the stream s. So the question is whether for other streams s' the generated $K^{s'}$ gives a different value for the same word w: $K^s(w) \neq K^{s'}(w)$. But it is easy to show that this cannot happen. First I show that

(*) for all words w_j^s and $w_k^{s'}$: if $w_j^s = w_k^{s'} = w$, then $K^s(w) = K^{s'}(w)$.

Assume otherwise, i.e., assume that $w_j^s = w_k^{s'} = w$, then $K^s(w) \neq K^{s'}(w)$. Let l be the first position on which $K^s(w) = ua\ldots$ and $K^{s'}(w) = ub\ldots$ differ, with $a \neq b$ being the symbols at position l. But then the $w_l^s \sqsubseteq w$ and $Q(w_l^s) = K^s(w_l^s) = ua$. On the other hand $w_l^{s'} \sqsubseteq w$ and $Q(w_l^{s'}) = K^{s'}(w_l^{s'}) = ub$. But either $w_l^{s'} \sqsubseteq w_l^s$ or $w_l^s \sqsubseteq w_l^{s'}$. In the first case $Q(w_l^s \circ s) = ub\ldots$ for any s due to the definition of $w_l^{s'}$. But $Q(w_l^s \circ s) = ua\ldots$ due to the definition of w_l^s which gives a contradiction. In the other $Q(w_l^{s'} \circ s) = ua\ldots$ for any s due to the definition of w_l^s, giving again a contradiction. This proves (*).

The only thing left to show is that no word w_i^s on stream s can be the same as a "non-aligned" $w_j^{s'}u$ on stream s' with $w_j^{s'} \sqsubset w_j^{s'}u \sqsubset w_{j+1}^{s'}$. But if $w_i^s = w_j^{s'}u$ were the case then w.lo.g we may assume that i is a minimal such one. But now for all $i' < i$ it can not be the case that $w_{i'}^s = w_{j'}^{s'}u'$ due to the minimality of i. So it must hold that $w_{i'}^s = w_{j'}^{s'}$ for some j'. Due to what was shown this leads to $K^s(w_{i'}^s) = K^{s'}(w_{j'}^{s'})$. Because $w_j^{s'} \sqsubset w_i^s$ we get that for $i' = i-1$ we have $w_j^{s'} = w_{i-1}^s$ and so $j = i-1$ or $j+1 = i$. But this cannot be the case as $w_{j+1}^{s'} \neq w_j^{s'}u$.

Proof of Proposition 5.5

By Theorem 5.3 Q is AC, hence the following chain of equations can be derived:

$$Q(s) = \bigcirc_{j=0}^{|s|} K(s^{\leq j}) = \bigcirc_{j=0}^{|s'|} K(s^{\leq j}) \circ \bigcirc_{k=|s'|+1}^{|s|} K(s^{\leq k})$$
$$= \bigcirc_{j=0}^{|s'|} K((s')^{\leq j}) \circ \bigcirc_{k=|s'|+1}^{|s|} K(s^{\leq k}) = Q(s') \circ \bigcirc_{k=|s'|+1}^{|s|} K(s^{\leq k})$$

so $Q(s') \sqsubseteq Q(s)$.

Proof of Proposition 5.6

Assume first that s is finite. The proof is by induction on length of s. Case $|s| = 0$, i.e., $s = \epsilon$. Then $s^{=0} = \epsilon$. Now assume $|s| = n + 1$, say $s = s'a$. Then $Q(s'a) = Q(s') \circ Q(a) = \bigcirc_{i=0}^{|s'|} Q((s')^{=i}) \circ Q(a) = \bigcirc_{i=0}^{|s|} Q(s^{=i})$. Now assume that s is infinite. We show that for all n-prefixes s" of $s = s'' \circ s'$ it is the case that $Q(s'' \circ s') = \bigcirc_{i=0}^{|s''|} Q((s'')^{=i}) \circ Q(s')$. But this is proven in the same way as for the finite case.

Now AC membership follows easily: one can set $K(\epsilon) = Q(\epsilon)$ and $K(wa) = Q(a)$ for any $w \in D^*$ and $a \in D$.

Proof of Proposition 5.10

1. $K(u \circ \epsilon) = K(u) = K(\epsilon)$ for all $u \in D^*$.
2. Let K be an i window. Let $|w| = j \geq i$ and $u \in D^*$. Then

$$K(uw) = K((uw^{=1} \dots w^{=j-i-1})w^{=j-i}w^{=j-i+1} \dots w^j)$$
$$= K(w^{=j-i}w^{=j-i+1} \dots w^j) = K((w^{=1} \dots w^{=j-i-1})w^{=j-i}w^{=j-i+1} \dots w^j)$$
$$= K(w)$$

3. One can set $K(\epsilon) = \epsilon$ and $K(wu) = Q(u)$ for all $w \in D^*, u \in D$. If, conversely, $K(\epsilon) = \epsilon$, then $Repeat(K)(\epsilon) = \epsilon$ and

$$Q(s \circ s') = Repeat(K)(s \circ s') = \bigcirc_{j=0}^{|s \circ s'|} K((s \circ s')^{\leq j})$$
$$= K(\epsilon) \circ \bigcirc_{j \geq 1}^{|s|} K((s \circ s')^{=j}) \circ \bigcirc_{j=|s|+1}^{|s \circ s'|} K((s \circ s')^{\leq j})$$
$$= \bigcirc_{j \geq 1}^{|s|} K((s)^{\leq j}) \circ \bigcirc_{j=1}^{|s'|} K((s')^{=j})$$
$$= \bigcirc_{j \geq 1}^{|s|} K((s)^{\leq j}) \circ \bigcirc_{j=1}^{|s'|} K((s')^{\leq j})$$
$$= Q(s) \circ Q(s')$$

Proof of Proposition 5.12

Let $s \in D^*$ and $|s| \geq 1$. Then, by induction, $Q(s)$ is a subsequence of s, hence also a finite word. So the first condition of (DISTRIBUTION) is fulfilled. Now let also $s' \in D^\infty$. Let $s = wu$ for some $w \in D^*$ and $u \in D$. Then $Q(s \circ s') = Q(wus')$. Then either $Q(s \circ s') = Q(wus') = Q(w) \circ u \circ Q(s')$. But one has also $Q(wu) = Q(w)u$ (taking s' in (TI-FILTER), so $Q(s \circ s') = Q(w) \circ u \circ Q(s') = Q(wu) \circ Q(s') = Q(s) \circ Q(s')$. Or, in the other case, $Q(s \circ s') = Q(wus') = Q(w) \circ Q(s')$. But again using (TI-FILTER) with $s' = \epsilon$ one also gets $Q(wu) = Q(w)$ and hence $Q(s \circ s') = Q(w) \circ Q(s') = Q(wu) \circ Q(s') = Q(s) \circ s'$.

Proof of Proposition 5.13

Assume $Q : D^\infty \longrightarrow D^\infty$ fulfils (FACTORING-N). If $|s| < n$, this is what is stated in (FACTORING-N). So $|s| \geq n$ and let $s = r \circ s'$ with $|r| = n$. Then $Q(r \circ s') = Q(r) \circ Q((r \circ s')^{\geq 2})$. On the right factor one can apply again the factorisation according to (FACTORING-N). Applying this repeatedly (using induction) gives the desired representation:

$$\begin{aligned}
Q(s) &= Q(r \circ s') \\
&= Q(r) \circ \underline{Q((r \circ s')^{\geq 2})} \\
&= Q(r) \circ Q((r \circ s')^{[2,n+1]}) \circ \underline{Q(((r \circ s')^{\geq 2}))^{\geq 3}} \\
&= \underline{Q(r)} \circ Q((r \circ s')^{[2,n+1]}) \circ Q(((r \circ s')^{\geq 3})) \\
&= \underline{Q(r^{[1,n]})} \circ \underline{Q((r \circ s')^{[2,n+1]})} \circ Q(((r \circ s')^{\geq 3})) \\
&= Q(s^{[1,n]}) \circ Q(s^{[2,n+1]}) \circ \underline{Q(((r \circ s')^{\geq 3}))} \\
&= Q(s^{[1,n]}) \circ Q(s^{[2,n+1]}) \circ Q(s^{[3,n+2]}) \circ Q(((r \circ s')^{\geq 4})) \\
&= \dots \\
&= Q(s^{[1,n]}) \circ Q(s^{[2,n+1]}) \circ Q(s^{[3,n+2]}) \circ \dots \circ Q(s^{[|s|-n,|s|]}) \\
&= \bigcirc_{j=1}^{|s|-n+1} Q(s^{[j,n+j-1]})
\end{aligned}$$

For infinite s the assertion follows for that from finite s.

Proof of Proposition 5.14

Let $n \geq 1$. Assume that Q fulfils (FACTORING-N). Define K by $K(w) = \epsilon$ for $|w| < n$. For $w \geq n$ let $K(w) = Q(w^{\geq |w|-n+1})$ = Q-value of n-suffix of w. Then $Q(s) = \bigcirc_{j=0}^{|w|} K(s^{\leq j}) = \epsilon$ for words s with $|s| < n$ by definition of K. For $s \geq n$ one has by proposition 5.13

$$Q(s) = \bigcirc_{j=1}^{|s|-n+1} Q(s^{[j,n+j-1]}) = \bigcirc_{j=1}^{|s|-n+1} K(s^{\leq n+j-1})$$
$$= \epsilon \circ \bigcirc_{j=n}^{|s|} K(s^{\leq j}) = \bigcirc_{j=1}^{n-1} K(s^{\leq j}) \circ \bigcirc_{j=n}^{|s|} K(s^{\leq j})$$

The other direction (namely, n-window induced functions fulfill (FACTORING-N) and monotonicity) is clear.

Proof of Theorem 5.24

Clearly, the range of each function f in MBINC is length-bounded, i.e., there is $m \in \mathbb{N}$ such that for all $w \in D^* : |f(w)| \leq m$. But then, according to [22, Proposition 22], f can be computed by a bounded-memory sASM. As the *Repeat* functional does (nearly) nothing else than the repeat-recursion rule, we get the desired representation.

The other direction is more advanced but can be mimicked as well: All basic rules, i.e., update rules can be modelled by ACCU functions (as one has to store only one symbol of the alphabet in each register; the update is implemented as accu-recursion). The parallel application is modelled by the parallel recursion principle in window-recursion. The if-construct can be simulated using cond. And the quantifier-free formula in the if construct can also be represented using cond as the latter is functionally complete.

Chapter 6
High-Level Declarative Stream Processing

Abstract High-level declarative stream processing deals with accessing streams via a declarative language such as a logic-based query language—in contrast to a procedural programming language, say. Moreover, the characterisation as high-level means that the semantics of such a language depends not only on the streams but also on a background knowledge base. This chapter discusses first-order logic rewritability results for STARQL, a query framework that is an instance of high-level declarative stream processing following the paradigm of ontology-based data access. Though STARQL provides user-friendly abstractions such as a sequencing mechanism for generating small aboxes from a stream, it can be shown that STARQL allows for first-order logic rewritability of queries.

6.1 Introduction

The former chapter presented foundational aspects of stream processing from the (infinite) word perspective, treating stream queries as functions from (infinite) words to (infinite) words, and developed representation results for various stream functions with complete axiomatic characterisations, i.e., representation results of the first category. In this chapter, the focus is on STARQL (Streaming and Temporal ontology Access with a Reasoning-based Query Language, pronounced Star-Q-L), a particular stream query language framework allowing to express stream queries in the sense above. As a corollary to investigations on two different semantics for STARQL, this chapter gives a representation result of the second category, namely, showing that STARQL queries can be rewritten into FOL queries.

STARQL is a *declarative language* in the sense that it provides a logic-based (instead, say, a procedural) access to streaming data, and it is a *high-level* query language in the sense that the semantics of the query language depends—next to the streams—on a static background knowledge base. In the main application scenarios, for which STARQL is intended to be used, the knowledge base is the union of an

Ö. L. Özçep, *Representation Theorems in Computer Science*,
https://doi.org/10.1007/978-3-030-25785-9_6

ontology (tbox) and data (abox). Hence, STARQL can be understood as a contribution to recent efforts on temporalizing and streamifying OBDA [3, 5, 9, 6, 32, 27].

Though in some aspects similar to other temporal and stream OBDA approaches mentioned above, STARQL differs mainly in the semantics of the window operator. Instead of loosely adapting the bag semantics for window operators in relational data stream managements (as done in many of the streamified OBDA approaches), STARQL has a window operator with a built-in sequencing strategy: at every evolving time point, the window produces a finite sequence of aboxes on which temporal reasoning can be applied. The motivation for adding a sequence strategy was born out of considerations on *adequate representations* for information needs over streams. Just using a bag-semantics, in which the timestamps of the stream elements are dropped, would have not allowed integrating the certain answer semantics for queries over a background tbox in an appropriate way.

The picture of relations between the different temporal and streamified OBDA approaches, in particular w.r.t. the semantics of specific logical operators such as the window operator, is still incomplete (a notable exception being the paper [4] which tries to define equivalences between stream programs.) To fill this gap, this chapter also gives an analysis of the query framework STARQL in comparison to other streaming and temporal OBDA approaches. In particular, it is shown that (a safe fragment of) TCQs [5] is embeddable into STARQL (Prop. 6.11).

This chapter is structured as follows. Section 6.2 describes the syntax and semantics of the STARQL framework. Section 6.3 introduces another semantics which fits better to a more standardised formulation of FOL rewritability and gives a proof of FOL rewritability for STARQL. Section 6.4 gives a comparison of the STARQL query language regarding its expressivity w.r.t. LTL like OBDA-based query languages, showing in particular how to embed the safe fragment of the temporal query language of TCQs into STARQL. Section 6.5 contains related work. The chapter is concluded by a summary in Sect. 6.6.

This chapter is mainly based on the publication [31], the only publication on STARQL with a clear reference to representation aspects. For other aspects of STARQL, I refer the reader to the papers that were published in the context of the Optique[1] project. A general overview of the Optique project (with a short high-level description of STARQL) can be found in the paper [14]. More details on the semantics of STARQL, its implementation within a stream engine, its optimisations, and further extensions can be found in the public deliverables [28, 24, 16, 29]. An early tutorial-style overview of STARQL is the main content of [30]. In [27], safety aspects of STARQL are discussed. Further recent developments w.r.t. STARQL towards an analytics-aware OBDA approach for use in industrial scenarios are given in [22, 18, 20, 21, 19]. Details on the implementation of the STARQL engines for historic reasoning can be found in [25, 26]. A visual interface for stream processing with STARQL is described in [34, 33].

[1] `optique-project.eu/`

6.2 The STARQL Framework

STARQL is a stream query language framework for OBDA scenarios with temporal and stream data. I call it a framework, because it describes a whole class of query languages which differ regarding the expressivity of the DL used for the tbox and regarding the embedded query languages used to query the individual intra-window aboxes constructed in the sequencing operation (see below).

6.2.1 Example

The following example for an information need in an agent scenario illustrates the main constructors of STARQL. A rational agent has different sensors, in particular different temperatures attached to different components. I assume that the agent receives both, high-level messages and low-level measurement messages, from a single input stream `Sin`. The agent has some background knowledge on the sensors. This knowledge is stored in a tbox. In particular, the tbox contains an axiom stating that all temperature sensors are sensors and that all type-X temperature sensors are temperature sensors. Factual knowledge on the sensors is stored in a (static) abox. For example, the abox may contain assertions *type-X-temperature-Sensor(tcc125), attachedTo(tcc125,c1), locatedAt(c1,rear)* stating that there is a temperature sensor of type X named $tcc125$ that is attached to some component $c1$ located at the rear of the physical agent. There is no explicit statement that $tcc125$ is a temperature sensor, this can be derived only with the axioms of the tbox.

The agent has to recognise whether the sensed temperature is critical. Due to some heuristics, a critical state is identified with the following pattern: in the last 5 minutes there was a monotonic increase on some interval followed by an alert message of category A, an A-message for short. The agent is expected to output every 10 second all temperature sensors showing this pattern and to mark them as critical. A STARQL formalisation of this information need is given in the listing of Fig. 6.1.

The `CONSTRUCT` operator (line 2) fixes the format of the output stream. Here, as well as in the `HAVING` clause (see below), STARQL uses the named-graph notation of the W3C recommended RDF[2] query language SPARQL[3] for fixing a basic graph pattern (BGP) and attaching a time expression. The output stream contains expressions of the form

```
GRAPH NOW { ?s a inCriticalState }
```

[2] `https://www.w3.org/RDF/`

[3] `https://www.w3.org/TR/rdf-sparql-query/`

```
1  CREATE STREAM Sout AS
2  CONSTRUCT  GRAPH NOW { ?s a inCriticalState }
3  FROM STREAM Sin[NOW-5min, NOW]->10s,  <http://abox>, <http://tbox>
4  USING PULSE AS START = 0s, FREQUENCY = 10s
5  WHERE { ?s a TempSens }
6  SEQUENCE BY StdSeq
7  HAVING
8  EXISTS i1, i2, i3
9   0 < i1 AND i2 < MAX AND plus(i2,1,i3)
10  GRAPH i3 { ?s :message ?m  . ?m a A-Message } AND
11  FORALL i, j, ?x,?y:
12    IF  i1 <= i  AND i<= j  AND j <= i2  AND
13        GRAPH i { ?s :hasVal ?x }  AND GRAPH j { ?s :hasVal ?y }
14    THEN ?x <= ?y
```

Fig. 6.1 Example STARQL query

where `NOW` is instantiated by time points and `?s` by constants fulfilling the required conditions as specified in the following lines of the query. The evolvement of the time `NOW` is specified in the pulse declaration (line 4).

The resources to which the query refers are specified using the keyword `FROM` (line 3). Following this keyword one can refer to one or more streams (by names or further stream expressions) and, optionally, by URIs to a tbox and an abox. In this example, only one stream is referenced, the stream named S_{in}. In this case, the stream consists, first, of timestamped triples matching the BGPs of the form

```
GRAPH t1 { ?s :hasVal ?y }
```

stating that `?s` has value `?y` at time `t1`. In logical notation, these subgraphs would be written as timestamped abox assertions of the form $hasVal(?s,?y)\langle t2\rangle$. Secondly, the input stream may contain timestamped BGPs of the form

```
GRAPH t2 { ?m a A-Message }
```

stating that at time point `t2` a message of type A arrived. In DL-notation this would be expressed as: $A\text{-}Message(?m)\langle t2\rangle$. The window operator attached to the input stream, `[NOW-5min, NOW]->10s`, is meant to give snapshots of the stream with the slide of 10s (update frequency) and range of 5 minutes (all stream elements within last 5 minutes).

The `WHERE` clause (line 5) specifies the sensors `?s` that the information need asks for, namely temperature sensors. It is clear that the agent has to incorporate his background knowledge from the tbox: in order to get all temperature sensors `?s` he has also to find all type-X sensors. The `WHERE` clause is evaluated only against the static abox. The stream-temporal conditions are specified in `HAVING` clause.

For every binding of `?s`, the query evaluates conditions that are specified in the `HAVING` clause (lines 7–14). A sequencing method (here `StdSeq`) maps an input stream to a sequence of aboxes (annotated by states i, j) according to a grouping criterion. The built-in sequencing method `StdSeq` is called *standard sequencing*. It puts all stream elements with the same timestamp into the same mini abox.

Testing for conditions at a state is done with the SPARQL sub-graph mechanism. So, e.g.,

`GRAPH i3 {?s :message ?m . ?m a A-Message}` (line 10)

asks whether `?s` showed a message of type A at state `i3`. State `i3` is further determined as the successor of the end state `i2` in the interval `[i1, i2]` (line 9). Over the interval `[i1, i2]` the usual monotonicity condition (`FORALL` condition, lines 11–14) is expressed using a first-order logic pattern.

Also in case of the `HAVING` clause, the background knowledge must be incorporated in order to guarantee a complete set of answers. For example, the tbox may contain a taxonomy of different types of messages, in particular different sub-A-type messages. If only instances of these subtypes are mentioned in the abox, then their super-types have to be inferred by the agent.

6.2.2 Syntax

The example in the previous subsection illustrated the syntax and the intended semantics of STARQL. In this section I describe a grammar that captures this example. This grammar leads to a sub-fragment of the original STARQL language [28]. In particular, the `HAVING` clauses are less expressive than the original ones. Further I left out aggregation constructors and macro definitions. For a full description I refer the reader to [28, 27].

The grammar (Fig. 6.2) is denoted STARQL(OL,ECL) and it contains parameters that have to be specified in its instantiations: the ontology language OL and the embedded condition language ECL. The ontology language OL constrains the languages of the aboxes and the tboxes that are referred in the grammar (underlined in Fig. 6.2). ECL is a query language referring to the signature of the ontology language. STARQL uses ECL conditions in its `WHERE` and `HAVING` clauses.

The adequate instantiation of STARQL(OL, ECL) may vary depending on the requirements of the use case. In Sect. 6.3 I consider the instantiation STARQL (DL-Lite, UCQ) with the prototypical ontology language/query language combination used in classical OBDA: DL-Lite/UCQs.

I am not going to discuss the whole grammar but only make some comments on the most interesting part, which is the set of rules for the specification of `HAVING` clauses (abbreviated *hCl* in the grammar). In the full STARQL grammar (see [27]) `HAVING` clauses are allowed to use arbitrary FOL constructors, in particular all boolean connectors, and also exists- as well as forall-quantifiers. As STARQL allows infinite domains (such as the real numbers in order to specify, say, temperature values) queries using FOL constructors have to be used with care in order to give safe queries, i.e., queries that output only finite sets of bindings. A query such as $\phi(y) = \neg hasVal(tcc125, y)$ for example is not safe as it would require outputting all of the infinitely many ys not being values of *tcc125*.

```
starql      → [prefix] createExp
createExp   → CREATE STREAM sName AS
              constrExp
pulseExp    → PULSE AS
              START = start,
              FREQUENCY = freq
constrExp   → CONSTRUCT cHead(x, y)
              FROM listWStrExp
              [, URIs – To – aboxes/tboxes]
              [USING pulseExp]
              [WHERE whereCl(x)]
              SEQUENCE BY seqMeth
              HAVING safeHCl(x, y)
cHead(x, y) → GRAPH timeExp triple(x, y)
              { . cHead(x, y)}
listWStrExp → (sName | constrExp) winExp
              [, listWStrExp]
winExp      → [timeExp1, timeExp2]->sl
timeExp     → NOW | NOW - constant

whereCl(x)        → ECL(x)
seqMeth           → StdSeq | seqMeth(~)
term(i)           → i
term()            → MAX | 0 | 1
arAt(i1, i2)      → term1(i1) op term2(i2)
                    (op ∈ {<,<=, =, >, >=})
arAt(i1, i2, i3)  → plus(term1(i1),
                         term2(i2),
                         term3(i3))
stateAt(x, i)     → GRAPH i ECL (x)
atom(x)           → arAt(x) | stateAt(x)
hCl(x)            → atom(x) | hCl(x) OR hCl(x)
hCl(x, y)         → hCl(x) AND hCl(y)
hCl(x)            → hCl(x) AND FORALL y
                    IF hCl(x, y) THEN hCl(x, y)
hCl(x, z)         → EXISTS y hCl(x, y) AND
                    hCl(z, y)
safeHCl(x)        → hCl(x)
                    (x contains no i variable)
```

Fig. 6.2 Syntax for STARQL(OL, ECL) template.

This problem is known since the beginning of classical DB theory and it has been handled by describing syntactical rules guaranteeing safeness. A similar approach for handling safeness, but relying on adornments, is described in [27]. The grammar presented here, which covers only a fragment of full STARQL, has no adornments but still reflects safety conditions. For example, the boolean connector for disjunction (or) is allowed to be applied only for disjunctions with the same set of open variables. (This point regarding disjunctions is discussed in more detail in Sect. 6.4). Furthermore, the existential and the forall-quantifiers are allowed to quantify only over variables which are guarded. Hence, an exists-quantifier over x is allowed only if x is bounded by a safe hCL clause appearing as conjunction in the scope of the exists quantifier. And universally bounded variables are allowed only if they are guarded as the antecedens of an implication in the scope of the forall-quantifier.

6.2.3 Semantics

The explication of the semantics for STARQL queries rests on the semantics of the instantiations of the parameter values OL and ECL. The only presumption I make is that OL and ECL have to fulfil the condition that there is a notion of a certain answer of an ECL w.r.t. an ontology. The motivation for such a layered—or as I called it here: separated—definition of the semantics is a strict separation of the semantics provided by the embedded condition languages ECL and the semantics used on top of it. Hence the separated semantics has a plug-in flavour, allowing users to embed any preferred ECL without repeatedly redefining the semantics of the whole query language.

For ease of exposition I assume that the query specifies only one output sub-graph pattern and that there is exactly one static abox $\mathcal{A}_{st}$ and one tbox $\mathcal{T}$. Similar to the approach of [5], the tbox is assumed to be non-temporal in the sense that there are no special temporal or stream constructors. I give a denotational specification $[\![S_{out}]\!]$ of S_{out} recursively by defining the denotations of the components. I will refer to the notion of a temporal abox within this denotation semantics and also later on. A *temporal abox* is a finite set of timestamped abox axioms $ax\langle t\rangle$, with $t \in T$. I call structures of the form $\langle(\mathcal{A}_i)_{i\in[n]}, \mathcal{T}\rangle$ consisting of a finite sequence of aboxes and a pure tbox a *sequenced ontology (SO)*. The index i of the abox $\mathcal{A}_i$ is called its *state*.

So assume that the following query template is given.

$$\begin{aligned} S_{out} = {} & \texttt{CONSTRUCT GRAPH } \mathit{timeExpCons}\ \Theta(\mathbf{x}, \mathbf{y}) \\ & \texttt{FROM } S_1\ \mathit{winExp}_1, \ldots, S_m\ \mathit{winExp}_m, \mathcal{A}_{st}, \mathcal{T} \\ & \texttt{WHERE } \psi(\mathbf{x})\ \texttt{SEQUENCE BY } \mathit{seqMeth}\ \texttt{HAVING } \phi(\mathbf{x}, \mathbf{y}) \end{aligned}$$

Windowing

Let $[\![S_i]\!]$ for $i \in [m]$ be the streams of timestamped abox assertions. The denotation of the windowed stream $ws_i = S_i\ [\mathit{timeExp}^i_1, \mathit{timeExp}^i_2]\text{->}sl_i$ is defined by specifying a function F^{winExp_i} s.t.: $[\![ws_i]\!] = F^{winExp_i}([\![S_i]\!])$.

The expression $[\![ws_i]\!]$ denotes a stream with timestamps from the set $T' \subseteq T$, where $T' = (t_j)_{j\in\mathbb{N}}$ is fixed by the pulse declaration with t_0 being the starting time point of the pulse. The domain of the resulting stream consists of temporal aboxes.

Assume that $\lambda t.g^i_1(t) = [\![\mathit{timeExp}^i_1]\!]$ and $\lambda t.g^i_2(t) = [\![\mathit{timeExp}^i_2]\!]$ are the unary functions of time denoted by the time expressions in the window. We have to define for every t_j the temporal abox $\tilde{\mathcal{A}}^i_{t_j}\langle t_j\rangle \in [\![ws_i]\!]$. If $t_j < sl - 1$, then $\tilde{\mathcal{A}}^i_{t_j} = \emptyset$. Else set first $t^i_{start} = \lfloor t_j/sl \rfloor \times sl$ and $t^i_{end} = max\{t_{start} - (g^i_2(t_j) - g^i_1(t_i)), 0\}$, and define on that basis $\tilde{\mathcal{A}}^i_{t_j} = \{ax\langle t\rangle \mid ax\langle t\rangle \in [\![S]\!] \text{ and } t^i_{end} \leq t \leq t^i_{start}\}$. Now, the denotations of all windowed streams are joined w.r.t. the timestamps in T': $js([\![ws_1]\!], \ldots, [\![ws_m]\!]) := \{(\bigcup_{i\in[m]} \tilde{\mathcal{A}}^i_t)\langle t\rangle \mid t \in T' \text{ and } \tilde{\mathcal{A}}^i_t\langle t\rangle \in [\![ws_i]\!]\}$.

Sequencing

The stream $js([\![ws_1]\!], \ldots, [\![ws_m]\!])$ is processed according to the sequencing method specified in the query. The output stream has timestamps from T'. The stream domain now consists of finite sequences of pure aboxes.

The sequencing methods used in STARQL refer to an equivalence relation $\sim$ to specify which assertions go into the same intra-window abox. The relation $\sim$ is required to respect the time ordering, i.e., it has to be a congruence over T. The equivalence classes are referred to as *states* and are denoted by variables i, j etc.

Let $\tilde{\mathcal{A}}_t\langle t\rangle$ be the temporal abox of $js([\![ws_1]\!], \ldots, [\![ws_m]\!])$ at t. Let $T'' = \{t_1, \ldots, t_l\}$ be the time points occurring in $\tilde{\mathcal{A}}_t$ and let k' the number of equivalence classes generated by the time points in T''. Then the sequence at t is defined as $(\mathcal{A}_0, \ldots, \mathcal{A}_{k'})$ where in turn for every $i \in [k']$ the abox $\mathcal{A}_i$ is defined be the following equation: $\mathcal{A}_i = \{ax\langle t'\rangle \mid ax\langle t'\rangle \in \tilde{\mathcal{A}}_t$ and t' in i^{th} equivalence class$\}$. The standard sequencing method `StdSeq` is just $seqMeth(=)$. Let $F^{seqMeth}$ be the function realizing the sequencing.

WHERE Clause

In the `WHERE` clause only $\mathcal{A}_{st}$ and $\mathcal{T}$ are relevant for the answers. So, purely static conditions (e.g. asking for sensor types as in the example above) are evaluated only on $\mathcal{A}_{st} \cup \mathcal{T}$. The result are bindings $\mathbf{a}_{wh} \in cert(\psi(\mathbf{x}), \langle \mathcal{A}_{st}, \mathcal{T}\rangle)$. This set of bindings is applied to the `HAVING` clause $\phi(\mathbf{x}, \mathbf{y})$.

HAVING Clause

STARQL's semantics for the `HAVING` clauses relies on the certain-answer semantics of the embedded ECL conditions.

The semantics of $\phi(\mathbf{a}_{wh}, \mathbf{y})$ has to be defined for every binding $\mathbf{a}_{wh}$ from the evaluation of the `WHERE` clause. For every t, one has to specify the bindings for $\mathbf{y}$. Assume that the sequence of aboxes at t is $seq = (\mathcal{A}_0, \ldots, \mathcal{A}_k)$. I define the set of *separation-based certain answers*, denoted: $cert_{sep}(\phi(\mathbf{a}_{wh}, \mathbf{y}), \langle \mathcal{A}_i \cup \mathcal{A}_{st}, \mathcal{T}\rangle)$.

If for any i the pure ontology $\langle \mathcal{A}_i \cup \mathcal{A}_{st}, \mathcal{T}\rangle$ is inconsistent, then we set $cert_{sep} =$ NIL, where NIL is a new constant not contained in the signature. In the other case, the bindings are defined as follows. For t one constructs a sorted first-order logic structure $\mathfrak{I}_t$: The domain of $\mathfrak{I}_t$ consists of the index set $\{0, \ldots, k\}$ as well as the set of all individual constants of the signature. For every state atom *stateAt* `GRAPH i` $ECL(\mathbf{z})$ in $\phi(\mathbf{a}_{wh}, \mathbf{y})$ with free variables $\mathbf{z}$ having length l, say, introduce an $(l+1)$-ary symbol R and replace `GRAPH i` $ECL(\mathbf{z})$ by $R(\mathbf{z}, i)$. The denotation of R in $\mathfrak{I}_t$ is then defined as the set of certain answers of the embedded condition $ECL(\mathbf{z})$ w.r.t. the ith abox $\mathcal{A}_i$: $R^{\mathfrak{I}_t} = \{(\mathbf{b}, i) \mid \mathbf{b} \in cert(ECL(\mathbf{z}), \langle \mathcal{A}_i \cup \mathcal{A}_{st}, \mathcal{T}\rangle)\}$. Constants denote themselves in $\mathfrak{I}_t$. This fixes a structure $\mathfrak{I}_t$ with finite denotations of its relation symbols. The evaluation of the `HAVING` clause is then nothing more than evaluating the FOL formula (after substitutions) on the structure $\mathfrak{I}_t$ (see Chap. 2 for the definition of satisfaction of FOL formulas).

Let $F^{\phi(\mathbf{a}_{wh}, \mathbf{y})}$ be the function that maps a stream of abox sequences to the set of bindings $(\mathbf{b}, t)$ where $\mathbf{b}$ is the binding for $\phi(\mathbf{a}_{wh}, \mathbf{y})$ at time point t.

Summing up, the following denotational decomposition results:

$$[\![S_{out}]\!] = \{\texttt{GRAPH}\ [\![timeExpCons]\!]\ \Theta(\mathbf{a_{wh}}, \mathbf{b}) \mid \mathbf{a_{wh}} \in cert(\psi(\mathbf{x}), \mathcal{A}_{st} \cup \mathcal{T}) \text{ and}$$
$$(\mathbf{b}, t) \in F^{\phi(\mathbf{a_{wh}}, \mathbf{y})}(F^{seqMeth}(js(F^{winExp_1}([\![S_1]\!]), \ldots, F^{winExp_m}([\![S_m]\!]))))\}$$

6.2.4 Properties of STARQL

Non-Reified Approach

A relevant question from the representational point of view is how to represent time in the query language. For STARQL, the decision was to use a non-reified approach, where time is handled in a special way. As illustrated by the agent example above, the abox assertions are tagged with timestamps. This method is similar to adding an extra time argument for concept and roles as in [3]. The non-reified approach allows representing time-dependent facts such as the fact that some sensor showed some value at a given time point. This time is relevant for the window semantics in STARQL.

An alternative strategy, following the idea of *reification*, is to handle time as an ordinary attribute attached to some entities ("res" in latin, hence the term "reification"). For example, in case of the agent scenario from the beginning of this chapter, it would be possible to describe temperature values by measurement objects. Then, the fact that a sensor shows a temperature value at some time could be expressed by stating that the measurement is associated with that sensor, has a timestamp with that time point, and has an associated value: $measurement(m_1)$, $hasSensor(m_1, s_0)$, $hasVal(m_1, 90°C)$, $hasTime(m_1, 30s)$. Here, the fact of a sensor showing a value at some time is expressed by some natural reification of the event, namely by introducing the measurements. Sometimes reification leads to objects which can be hardly coined natural (see, e.g., the discussion in [13]).

As the reified approach is more conservative and does not require to change the semantics, a natural question is whether STARQL follows the non-reified strategy. The main reason is that time requires a special treatment as it has specific constraints for reasoning. For example, in the measurement scenario one would like to express the time-dependent constraint that, at every time point, a sensor shows at most one value. This can be done with a classical DL-Lite axiom by stating $(func\ val) \in \mathcal{T}$. Note that under such a constraint it is necessary that the window semantics preserves the timestamps, as is indeed the case for the STARQL window semantics. Otherwise two timestamped stream elements of the form $hasVal(tcc125, 92°)\langle 3s\rangle$ and of the form $hasVal(tcc125, 95°)\langle 5s\rangle$ stating would lead to an inconsistency.

On the other hand, if one follows the reified approach such a time-dependent constraint is not expressible in a DL: one would have to formulate that there are no two measurements with the same associated sensor and same timestamp but different values. As DLs are concept-oriented, they cannot express such a non-tree shaped constraint.

Homogeneous Interface

For the syntax and the semantics of STARQL queries the exact resource of the input stream is not relevant: it may be a stream of elements arriving in real-time via a TCP port, but equally a simulated stream of data produced by reading out a text file or a temporal database. In the former case, one can speak of (genuine) stream querying, whereas in the latter case I use the term *historic querying*. So STARQL offers the same interface to real-time queries (as required, for example, in monitoring scenarios) and historic queries (as required, e.g., in reactive diagnostics). And, indeed such a homogenous interface to two different modes of querying has proved useful for real industrial use cases, in particular, for the turbine-diagnostics use case of SIEMENS in the context of the Optique project [22, 20, 21].

Separation between Static and Temporal Conditions

As illustrated in the example above, STARQL allows to separate the conditions expressed in an information need into conditions that concern only the static part of the background knowledge (tbox $\mathcal{T}$ and static abox $\mathcal{A}_{st}$) and into conditions which require both, the static part and the streams. The former can be queried in the `WHERE` clause, the latter in the `HAVING` clause. Looking at the semantics of `HAVING` clause, one sees that the `HAVING` clause also refers to the tbox and the static abox. And indeed, this reference, at first sight, is not eliminable. The reference to the tbox can be eliminated by just rewriting the `HAVING` clause into a new `HAVING` clause using the standard perfect-rewriting technique. Still, the theoretical question remains whether it is possible to push all references to the static abox (all occurrences of concept and role symbols that appear in the static abox) into the `WHERE` clause, so that the `HAVING` clause can be evaluated only on the streams and the bindings resulting from the evaluation of the `WHERE` clause. In other terms, is the `HAVING` clause separable in a pure static part and a part containing only role and concept symbols not part of the static abox? This is an open problem.

Even if separability in the sense above holds, in terms of feasible implementation, the reference to a large static abox remains a challenging problem. As of the time of writing up this monograph, and as far as I know, this problem has not been solved satisfactorily by any of the current temporal and streamified OBDA systems.

An Alternative Operational Semantics

The window semantics defined above is denotational and mimics the window operator definitions for CQL [2], which is one of the first relational data stream query languages. From the implementation point of view, an operational semantics is more helpful than a denotational semantics—at least it gives a different perspective on the intended semantics of the window. Furthermore, the operational view also sheds light on why the window definition was chosen exactly the way as stated above. The operational semantics is illustrated with the query template given in the listing of Fig. 6.3.

```
CREATE STREAM S_{out}

...
FROM Sin [NOW-wr, NOW] -> sl
USING PULSE WITH START = st, FREQUENCY = fr
...
```

Fig. 6.3 Example template query for illustration of operational semantics

Referring to the STARQL grammar, Fig. 6.3 instantiates $timeExp_1$ = `NOW-wr`, where `wr` is a constant denoting the window range, and $timeExp_2$ = `NOW`. I distinguish between a pulse time t_{pulse} and a stream time t_{str}. (For more than one stream one would have more local stream times.) The pulse time t_{pulse} evolves regularly according to the frequency specification,

$$t_{pulse} = st \longrightarrow st + fr \longrightarrow st + 2fr \longrightarrow \dots$$

In contrast, the stream time t_{str} is jumping/sliding and is determined by the trace of endpoints of the sliding window. More concretely, the evolvement of t_{str} is specified as follows:

$$t_{str} \xrightarrow{\text{IF } t_{str} + m \times sl \leq t_{pulse} \text{ (for } m \in \mathbb{N} \text{ maximal)}} t_{str} + m \times sl$$

The window contents at t_{pulse} is given by: $\{ax\langle t\rangle \in S_{in} \mid t_{str} - wr \leq t \leq t_{str}\}$. Note that always $t_{str} \leq t_{pulse}$. This a crucial point regarding the homogeneity aspect mentioned above: having always $t_{str} \leq t_{pulse}$ guarantees that applying the window on real-time streams does not give different stream elements than applying the window on a simulated stream from a DB with historic data. Otherwise, i.e., if $t_{str} > t_{pulse}$, the window in a historic query would contain future elements from $[t_{pulse}, t_{str}]$ whereas in the real-time case the window cannot contain future elements from $[t_{pulse}, t_{str}]$.

As an example, consider the STARQL in the listing of Fig. 6.4. Then one gets the

```
CREATE STREAM Sout AS
...
FROM STREAM Sin:  [NOW-3s, NOW] -> "3S"^^xsd:duration
USING PULSE WITH START =  "2005-01-01T00:00:00CET"^^xsd:dateTime,
                 FREQUENCY = "2S"^^xsd:duration
...
```

Fig. 6.4 Example query for illustration of operational semantics on one stream

following evolvements of the pulse time and the streaming time.

$$t_{pulse} : 0s \to 2s \to 4s \to 6s \to 8s \to 10s \to 12s \to$$
$$t_{str} : 0s \to 0s \to 3s \to 6s \to 6s \to 9s \to 12s \to$$

The example query in the listing of Fig. 6.5 refers to multiple streams and is intended to illustrate the synchronisation effect of the pulse:

$$t_{pulse} : 0s \to 2s \to 4s \to 6s \to 8s \to 10s \to 12s \to$$
$$t_{str_1} : 0s \to 0s \to 3s \to 6s \to 6s \to 9s \to 12s \to$$
$$t_{str_2} : 0s \to 2s \to 4s \to 6s \to 8s \to 10s \to 12s \to$$

Synchronicity

Similar to the semantics defined for the relational stream query language CQL [2], I assumed for the semantics of STARQL synchronised streams., i.e., input streams where the timestamps respect the arrival order. It is possible to defer the handling of asynchronous streams to the implementation level as done by [2]. And also this is foreseen in the layered approach of [23], where all synchronisation is handled on the levels below. But this is not a must when designing a stream processing system on the abstraction level of ontologies. Regarding the synchronicity aspect, e.g., it is a matter of flexibility to give also the user of the ontological query language a means to specify the way he wants to handle asynchronous streams directly, and even doing the specification for each stream query—independently of the other queries. A possibility for this is the use of a slack parameter. And, in fact, STARQL as defined in the technical report [28] is intended to handle also these.

6.2.5 Rewritability of HAVING Clauses

From the representational point of view, the most important property of STARQL queries is that its HAVING clauses are FOL rewritable under the separation-based semantics. In order to state the proposition, the rewritability notion has to be adapted to the case of a sequential ontology. I suggest the following natural notion of rewritability: Let $O = \langle (\mathcal{A}_i)_{i \in [n]}, \mathcal{T} \rangle$ be an SO. The canonical model $DB((\mathcal{A}_i)_{i \in [n]})$ of a

```
CREATE STREAM Sout AS
CONSTRUCT { ?sens rdf:type :RecentMonInc }<NOW>
FROM  STREAM Sin1 [NOW-3s, NOW]->"3S"^^xsd:duration,
      STREAM Sin2 [NOW-3s, NOW]->"2S"^^xsd:duration
USING  PULSE WITH START   = "2005-01-01T00:00:00CET"^^xsd:dateTime,
                  FREQUENCY = "2S"^^xsd:duration
SEQUENCE BY StdSeq AS seq
HAVING  (...)
```

Fig. 6.5 Example query for illustration of operational semantics on two streams

sequence of aboxes is defined as the sequence of minimal Herbrand models $DB(\mathcal{A}_i)$ of the aboxes $\mathcal{A}_i$. Let QL_1 and QL_2 be two query languages over the same signature of an SO and let OL be a language for the sequenced ontologies SO.

Definition 6.1 QL_1 allows for QL_2 rewriting of query answering w.r.t. ontology language OL iff for all queries ϕ in QL_1 and tboxes $\mathcal{T}$ in OL there exists a query $\phi_{\mathcal{T}}$ in QL_2 such that for all $n \in \mathbb{N}$ and all sequences of aboxes $(\mathcal{A}_i)_{i\in[n]}$ it holds that: $cert(\phi, \langle(\mathcal{A}_i)_{i\in[n]}, \mathcal{T}\rangle) = ans(\phi_{\mathcal{T}}, DB(\mathcal{A}_i)_{i\in[n]}))$

I call an ECL language *rewritability closed* w.r.t. OL iff ECL allows for perfect rewriting such that the rewritten formula is again an ECL condition.

Proposition 6.2 *Let OL be an ontology language and let ECL be a rewritability-closed condition language w.r.t. OL. Denote by QL_1 the instantiation of* `HAVING` *clauses with this ECL and OL. Then QL_1 allows for QL_1 rewriting of (separation-based) certain query answering w.r.t. OL.*

This is an immediate consequence of the separated semantics.

6.3 Separation-Based versus Holistic Semantics

This section introduces a new semantics for STARQL `HAVING` clauses, called *holistic* semantics, and shows that for a fragment of the `HAVING` clauses the holistic and the separated semantics are actually the same. One reason for introducing a holistic semantics is that it allows to compare the expressivity of STARQL with temporal-logic oriented query languages. In particular, I show that the `HAVING` fragment captures (a safe fragment) of the LTL-inspired query language of temporal conjunctive queries (TCQs) [5].

I assume that I have a `HAVING` clause where the free variables $\mathbf{x}$ of the `WHERE` clause are instantiated already and hence can be considered to be bounded.

With every time point t an SO $\langle(\mathcal{A}_i)_{i\in[n_t]}, \mathcal{T}\rangle$ is associated. The length of the sequence n_t may depend on the time point t. As I now fix the length, I write just n for n_t.

Definition 6.3 Let $O = \langle(\mathcal{A}_i)_{i\in[n_t]}, \mathcal{T}\rangle$ be an SO. Let $\hat{\mathfrak{I}} = (\mathfrak{I}_i)_{i\in[n]}$ be a sequence of interpretations $\mathfrak{I}_i = (\Delta, \cdot^{\mathfrak{I}_i})$ over a fixed non-empty domain Δ where the constants' interpretations do not change for different $\mathfrak{I}_i, \mathfrak{I}_j$. Then $\hat{\mathfrak{I}}$ is a model of O (written $\hat{\mathfrak{I}} \models O$) iff $\mathfrak{I}_i \models \langle\mathcal{A}_i, \mathcal{T}\rangle$ for all $i \in [n]$.

All interpretations $\mathfrak{I}_i$ have the same domain Δ. The constants' denotations does not change from state to state, hence they are considered rigid.

The tbox $\mathcal{T}$ is assumed to be a non-temporal tbox. Its inclusions hold at every time point. For example, a tbox axiom $temperatureSensor \sqsubseteq Sensor$ means that at every time point a temperature sensor is a sensor.

One can easily verify that the set of models of $O = \langle(A_i)_{i\in[n]}, \mathcal{T}\rangle$ is just the cartesian product of all sets of models $\mathfrak{I}_i$ of the local ontologies $\mathcal{T} \cup \mathcal{A}_i$.

Proposition 6.4 $\{\hat{\mathfrak{I}} \mid \hat{\mathfrak{I}} \models \langle (A_i)_{i\in[n]}, \mathcal{T}\rangle\} = \times_{i\in[n]}\{\mathfrak{I} \mid \mathfrak{I} \models \langle A_i, \mathcal{T}\rangle\}$

The satisfaction relation between sequences of interpretations and Boolean `HAVING` clauses is defined as follows.

Definition 6.5 Let $\hat{\mathfrak{I}} = (\mathfrak{I}_i)_{i\in[n]}$ be a sequence of interpretations and ν be an assignment of individuals $d \in \Delta$ to individual variables and numbers $i \in [n]$ to state variables i, j. Let $\nu[j \mapsto \underline{j}]$ be the variant of ν where j is assigned $\underline{j}$. The semantics are defined as follows:

$$
\begin{aligned}
\hat{\mathfrak{I}}, \nu \models \texttt{GRAPH i } \alpha &\text{ iff } \mathfrak{I}_{\nu(i)} \models \alpha \\
\hat{\mathfrak{I}}, \nu \models \texttt{EXISTS i } \phi &\text{ iff } \text{there is } \underline{i} \in [n] \text{ s.t. } \hat{\mathfrak{I}}, \nu[i \mapsto \underline{i}] \models \phi \\
\hat{\mathfrak{I}}, \nu \models \texttt{FORALL i } \phi &\text{ iff } \text{for all } \underline{i} \in [n] \text{ it holds that } \hat{\mathfrak{I}}, \nu[i \mapsto \underline{i}] \models \phi \\
\hat{\mathfrak{I}}, \nu \models \texttt{EXISTS x } \phi &\text{ iff } \text{there is } d \in \Delta \text{ s.t. } \hat{\mathfrak{I}}, \nu[x \mapsto d] \models \phi \\
\hat{\mathfrak{I}}, \nu \models \texttt{FORALL x } \phi &\text{ iff } \text{for all } d \in \Delta \text{ it holds that } \hat{\mathfrak{I}}, \nu[x \mapsto d] \models \phi \\
\hat{\mathfrak{I}}, \nu \models \phi_1 \texttt{ AND } \phi_2 &\text{ iff } \hat{\mathfrak{I}}, \nu \models \phi_1 \text{ and } \hat{\mathfrak{I}}, \nu \models \phi_2 \\
\hat{\mathfrak{I}}, \nu \models \phi_1 \texttt{ OR } \phi_2 &\text{ iff } \hat{\mathfrak{I}}, \nu \models \phi_1 \text{ or } \hat{\mathfrak{I}}, \nu \models \phi_2 \\
\hat{\mathfrak{I}}, \nu \models \phi_{arAt} &\text{ iff } \mathfrak{I}_0, \nu \models \phi_{arAt} \\
\hat{\mathfrak{I}} \models \phi &\text{ iff } \hat{\mathfrak{I}}, \emptyset \models \phi
\end{aligned}
$$

For arbitrary `HAVING` clauses $\phi(\mathbf{x})$ and a sequence of interpretations $\hat{\mathfrak{I}}$, the set of answers is defined as $ans(\phi(\mathbf{x}), \hat{\mathfrak{I}}) = \{\mathbf{a} \mid \hat{\mathfrak{I}} \models \phi(\mathbf{a})\}$. The set of certain answers w.r.t. an SO $\mathcal{O} = \langle (A_i)_{i\in[n]}, \mathcal{T}\rangle$ is: $cert_h(\phi, \mathcal{O}) = \bigcap_{\hat{\mathfrak{I}} \models \mathcal{O}} ans(\phi, \hat{\mathfrak{I}})$

I consider a fragment of the `HAVING` clauses for OL = DL-Lite and ECL = UCQ and denote it by $\mathcal{L}^{\exists}_{HCL}$: In this fragment I disallow the operators `FORALL x` and `EXISTS x`. In contrast, the implicit existential quantifiers in the UCQs are allowed.

I show that the original separation-based semantics and the holistic certain answer semantics (denoted $cert_h$) are the same on this fragment.

Theorem 6.6 *For any SO* $\mathcal{O} = \langle (A_i)_{i\in[n]}, \mathcal{T}\rangle$ *and any* $\phi \in \mathcal{L}^{\exists}_{HCL}$ *the following equality holds:* $cert_h(\phi, \mathcal{O}) = cert_{sep}(\phi, \mathcal{O})$.

Proof See page 149. □

As a corollary to the theorem above the rewritability for STARQL queries in the holistic semantics results.

Theorem 6.7 *Let* QL_1 *be the instantiation of the* `HAVING` *clause language with ECL = UCQ and OL = DL-Lite. Then* QL_1 *allows for* QL_1 *rewriting for holistic certain query answering w.r.t. OL on the abox sequence.*

Moreover, as `HAVING` clauses are based on FOL with $<$, `plus` (the `MAX` operator can be rewritten with $<$) and the state variables i can be pushed into the UCQs, one gets the following additional corollary:

Corollary 6.8 *Let QL_1 be the instantiation of* `HAVING` *clauses with ECL = UCQ and OL = DL-Lite. Then QL_1 allows for $FOL(<, \texttt{plus})$ rewriting for holistic certain query answering w.r.t. OL on the abox sequence.*

These rewritability theorems still do not say whether it is required to materialise the abox sequence or whether it can be defined as a view. In the end, the rewriting given above is "local" in the sense that it concerns only the sequential ontology within a window. If a rewriting w.r.t. the whole stream is required, then one has to incorporate the sequencing strategy, the pulse, and the window semantics.

The pulse function defines the outputs of the STARQL query. So the rewritten query must be informed about the time points. I formalise this by allowing the FOL query to refer to just the list of time points. The window contents can be calculated with some simple arithmetics using the order $<$, the addition operator and a multiplication operator. For the standard sequencing method there is a one-to-one correspondence between pairs of state and developing time (i, NOW) and timestamps which can be expressed easily as a FOL query with some arithmetic incorporating a multiplication operation `mult`. So one gets the following result:

Corollary 6.9 *Let QL_1 be the instantiation of* `HAVING` *clauses with ECL = UCQ and OL = DL-Lite and assume that the sequence of aboxes is generated by standard sequencing. Then QL_1 allows for $FOL(<, \texttt{plus}, \texttt{mult})$ rewriting for holistic certain query answering w.r.t. OL on the stream of timestamped abox axioms.*

With this corollary on the rewritability of STARQL's `HAVING` clauses the main step has been taken in order to show that STARQL queries (as a whole) are transformable in relational data stream languages such as CQL [2]. For this transformation also the specification of an unfolding mechanism is required which I have not discussed here (see, e.g., [26].)

6.4 Comparison with TCQs

The fragment of the `HAVING` clause language, for which I showed the equivalence of the holistic and the separation-based semantics, is still expressive enough to simulate query languages such as the language of temporal conjunctive queries (TCQs) [5], that combine temporal-logic operators with lightweight DL languages.

TCQs are defined by following a weak integration of conjunctive queries (CQs) and a linear temporal logic (LTL) template.

The syntax is given on the left-hand side in Fig. 6.6, the semantics according to the following definition and the right-hand side in Fig. 6.6.

Definition 6.10 Let ϕ be a Boolean TCQ. For $\hat{\mathfrak{I}} = (\mathfrak{I}_i)_{i \in [n]}$ and $i \in [n]$ one defines $\hat{\mathfrak{I}}, i \models \phi$ by induction on the structure of ϕ as in the right-hand side of Fig. 6.6.

The set of answers at i is defined as $ans(\phi, \hat{\mathfrak{I}}, i) = \{\mathbf{a} \mid \hat{\mathfrak{I}}, i \models \phi(\mathbf{a})\}$. For a SO O the set of certain answers at i is $cert(\phi, O) = \bigcap_{\hat{\mathfrak{I}} \models \mathcal{I}} ans(\phi, \hat{\mathfrak{I}}, i)$. The set of (certain)

$$
\begin{array}{l}
\phi \longrightarrow CQ \mid \\
\quad \phi \wedge \phi \mid \phi \vee \phi \mid \\
\quad \bigcirc\phi \mid \bullet\phi \mid \\
\quad \bigcirc^{-}\phi \mid \bullet^{-}\phi \mid \\
\quad \phi \,\mathsf{U}\, \phi \mid \phi \,\mathsf{S}\, \phi
\end{array}
\qquad
\begin{array}{rll}
\hat{\mathfrak{I}}, i \models \exists y_1, \ldots, y_m.\rho & \text{iff} & \mathfrak{I}_i \models \exists y_1, \ldots, y_m.\rho \\
\hat{\mathfrak{I}}, i \models \phi_1 \wedge \phi_2 & \text{iff} & \hat{\mathfrak{I}}, i \models \phi_1 \text{ and } \hat{\mathfrak{I}}, i \models \phi_2 \\
\hat{\mathfrak{I}}, i \models \phi_1 \vee \phi_2 & \text{iff} & \hat{\mathfrak{I}}, i \models \phi_1 \text{ or } \hat{\mathfrak{I}}, i \models \phi_2 \\
\hat{\mathfrak{I}}, i \models \bigcirc\phi_1 & \text{iff} & i < n \text{ and } \hat{\mathfrak{I}}, i+1 \models \phi_1 \\
\hat{\mathfrak{I}}, i \models \bullet\phi_1 & \text{iff} & i < n \text{ implies } \hat{\mathfrak{I}}, i+1 \models \phi_1 \\
\hat{\mathfrak{I}}, i \models \bigcirc^{-}\phi_1 & \text{iff} & i > 0 \text{ and } \hat{\mathfrak{I}}, i-1 \models \phi_1 \\
\hat{\mathfrak{I}}, i \models \bullet^{-}\phi_1 & \text{iff} & i > 0 \text{ implies } \hat{\mathfrak{I}}, i-1 \models \phi_1 \\
\hat{\mathfrak{I}}, i \models \phi_1 \,\mathsf{U}\, \phi_2 & \text{iff} & \exists k : i \leq k \leq n, \hat{\mathfrak{I}}, k \models \phi_2 \\
 & & \text{and } \hat{\mathfrak{I}}, j \models \phi_1 \forall j, i \leq j < k \\
\hat{\mathfrak{I}}, i \models \phi_1 \,\mathsf{S}\, \phi_2 & \text{iff} & \exists k : 0 \leq k \leq i, \hat{\mathfrak{I}}, k \models \phi_2 \\
 & & \text{and } \hat{\mathfrak{I}}, j \models \phi_1 \forall j, k < j \leq i.
\end{array}
$$

Fig. 6.6 TCQs: syntax and semantics

answers is defined as the (certain) answers at the last state: $ans(\phi, \hat{\mathfrak{I}}) = ans(\phi, \hat{\mathfrak{I}}, n)$ and $cert(\phi, O) := cert(\phi, O, n)$.

A simple example shows that TCQs are not domain independent—though it is intended to be used in the OBDA paradigm: Take $\phi(x, y) = A(x) \vee B(y)$ and consider interpretations $\mathfrak{I} = (\{a\}, \cdot^{\mathfrak{I}}), \mathfrak{J} = (\{a, b\}, \cdot^{\mathfrak{J}})$ with $A^{\mathfrak{I}} = A^{\mathfrak{J}} = \{a\}$ and $B^{\mathfrak{I}} = B^{\mathfrak{J}} = \emptyset$: one has $(a, b) \in ans(\phi, \mathfrak{J})$ but $(a, b) \notin ans(\phi, \mathfrak{I})$. As the TCQs are not domain independent, I consider the following safe fragment TCQ^s where the rule for $\vee$ is replaced by: if ϕ_1, ϕ_2 are in TCQ^s and have the same free variables, then $\phi_1 \vee \phi_2$ is in TCQ^s; $\bullet$, $\bullet^{-}$ are disallowed; the rules for S, U are replaced by: if ϕ_1, ϕ_2 are in TCQ^s and have the same free variables, then $\phi_1 \,\mathsf{U}\, \phi_2$ and $\phi_1 \,\mathsf{S}\, \phi_2$ are in TCQ^s.

It can be shown that TCQ^{s}s are embeddable into STARQL `HAVING` clause. The translation θ is the usual one known from translating modal logics (and description logics) into predicate logic. It just simulates the semantics for TCQs within the object language of STARQL. For example, $\theta_i(\exists y_1, \ldots, y_m.\psi)$ = `GRAPH i` ψ; $\theta_i(\phi_1 \vee \phi_2)$ = $\theta_i(\phi_1)$ `OR` $\theta_i(\phi_2)$; $\theta_i(\bigcirc^{-}\phi_1) = i > 0$ `AND` $\theta_{i-1}(\phi_1)$. From the similarity of the semantics for TCQ^s and the holistic semantics of $\mathcal{L}^{\exists}_{HCL}$-queries one can see that the transformation θ yields for every TCQ^s ϕ a $\mathcal{L}^{\exists}_{HCL}$-query with the same set of certain answers.

Proposition 6.11 *For all SOs O and TCQ^s ϕ: $cert(\phi, O) = cert_h(\theta(\phi), O)$*

Due to some unsafe operators in TCQs, STARQL embeds not all TCQs. This is an intended feature of STARQL as it was meant to be applicable for strict OBDA scenarios where safe query languages such as SQL are used for the backend sources.

The full language of TCQs allows for queries for which an implementation in a domain independent language such as SQL leads to performance issues: in the case of arbitrary disjunctions a function has to be implemented that returns all constants in the active domain. On the other hand, STARQL is more expressive than the safe fragment of TCQs: it offers (different) means to generate abox sequences, whereas for TCQs it is assumed that these are given in advance. Moreover, TCQs allow for quantifiers only within embedded CQs, but cannot handle outer quantification, which is needed in order to, e.g., express the monotonicity condition as in the example from the beginning.

In this section I considered only the expressivity of the `HAVING` fragment in comparison to the LTL inspired framework of TCQs. A comparison of the whole STARQL query language with other stream-temporal query languages can be found in the STARQL chapter of [21]. The main results of this comparison is that STARQL can compete with other stream-temporal languages w.r.t. the offered expressivity (though regular path queries as required in SPARQL 1.1[4] are not supported) and that it has a unique standing w.r.t. the support of a flexible sequencing mechanism.

6.5 Related Work

There are various, quite distinct approaches to high-level declarative stream processing, be it the work on relational data stream managements [2, 7, 15, 23], on SPARQL-derived stream-query languages [9, 6, 32] or on complex event processing [1, 8]. For some recent approaches the reader is also referred to papers published in the proceedings of the workshop on high-level declarative stream processing, Hidest'15[5] and Hidest'18[6].

But, investigations regarding representational issues in the realm of high-level temporal and stream processing are rather rare. They appear implicitly in investigations on FOL rewritability such as the framework of TCQs [5], which was extensively discussed here, or such as the framework of [3] which considers tboxes with temporal operators.

From a more abstract and sociological point of view, the efforts in finding a standard for high-level declarative query languages over temporal and stream data can be understood as a search for the essential concepts, the hidden structures underlying all current stream-temporal query languages. Comparisons on the expressivity of different stream query languages—either practically by benchmarks and experimental setups as in [11] or theoretically as in [4]—can be understood as the search for the structure preserving mappings according to the representation scenario framework of this monograph. It should be noted, that the various communities of stream researchers have not converged to such a standard. This is true, e.g., for the community on RDF stream reasoning[7] (see [10, 12] as a first try) but also for the older community working on relational data stream management systems ([17] can be understood as a first try).

---

[4] `https://www.w3.org/TR/sparql11-query/`

[5] `https://www.ifis.uni-luebeck.de/hidest15.html`

[6] `https://lat.inf.tu-dresden.de/~koopmann/HiDeSt18/`

[7] `https://www.w3.org/community/rsp/`

6.6 Summary

The STARQL query language framework is part of the recent venture of adapting classical OBDA for stream-temporal reasoning. As explained in the beginning of the monograph, the OBDA paradigm has a strong representational aspect in form of FOL rewritability. In this chapter I developed a representation result along this line by showing a rewritability result for STARQL `HAVING` clauses. Further representational issues came up with the correct representation of time (reified vs. non-reified).

The rewritability results shown in this chapter are trivial for the separation-based semantics and are a little bit harder for the holistic semantics of STARQL queries. Rewritability becomes even harder if one allows the tbox to incorporate temporal operators as in [3]. The focus on classical tboxes, as assumed in this chapter, is a clear restriction in expressivity which may lead to undesirable consequences in diagnostic scenarios. For example, consider the following tbox

$$\{\exists tempVal \sqsubseteq TempSens,\ \exists pressVal \sqsubseteq PressSens,\ TempSens \sqsubseteq \neg PressSens\}$$

and assume that the input stream is

$$S_{in} = \{\ldots tempVal(s0, 90)\langle 3s\rangle, pressVal(s0, 70)\langle 4s\rangle \ldots\}$$

The information regarding the sensor $s0$ is not consistent with the tbox, as it would entail that $s0$ is both a pressure sensor and a temperature sensor, which is excluded by the tbox. This inconsistency will not be detected if the static abox does not know about $s0$ and hence does not classify it. If the tbox were allowed to use temporal constructors, then one could express necessary rigidity assumptions. For example, one would state that if a sensor is a temperature sensor at some time, then it will be in all future time points (and was in all past time points).

As a stream-temporal OBDA query language allowing for rewritability (and also unfoldability), STARQL is a candidate for applications requiring a declarative conceptualisation over temporal and stream data. Next to the obvious scenarios for query answering in a central knowledge base, also high-level stream processing within a rational agent (see introduction of this monograph) are potential application scenarios. In rational agents various transformation tasks from low-level sensor stream data to high-level declarative stream data are required. In a classical OBDA approach these transformations would not have to be actually implemented, rather the high-level query on the virtual high-level stream would be transformed to a low-level query on the low-level sensor streams. As far as I know, there are no agent systems applying OBDA for the transformations mentioned above.

In this monograph, I spared out completely implementation aspects for STARQL. Descriptions of STARQL implementations can be found in the Optique deliverables [24, 16, 29] and in the papers [25, 26]. From a theoretical point of view, the most interesting question is to invent a good strategy of combining the mini-aboxes of the window with the possibly large static abox. In general, considering the moderate query answering times of current high-level stream engines on large static background knowledge, a good theory for a combination of streaming and large static data is still lacking.

References

1. Anicic, D., Rudolph, S., Fodor, P., Stojanovic, N.: Stream reasoning and complex event processing in ETALIS. Semantic Web **3**(4), 397–407 (2012)
2. Arasu, A., Babu, S., Widom, J.: The CQL continuous query language: Semantic foundations and query execution. The VLDB Journal **15**, 121–142 (2006)
3. Artale, A., Kontchakov, R., Wolter, F., Zakharyaschev, M.: Temporal description logic for ontology-based data access. In: Proceedings of the 23rd International Joint Conference on Artificial Intelligence (IJCAI-13), pp. 711–717 (2013)
4. Beck, H., Dao-Tran, M., Eiter, T.: Equivalent stream reasoning programs. In: S. Kambhampati (ed.) Proceedings of the 25th International Joint Conference on Artificial Intelligence (IJCAI-16), pp. 929–935. IJCAI/AAAI Press (2016)
5. Borgwardt, S., Lippmann, M., Thost, V.: Temporal query answering in the description logic DL-Lite. In: Proceedings of the 8th International Symposium on Frontiers of Combining Systems (FroCos-13), *LNCS*, vol. 8152, pp. 165–180 (2013)
6. Calbimonte, J.P., Jeung, H., Corcho, O., Aberer, K.: Enabling query technologies for the semantic sensor web. International Journal of Semantic Web Information Systems **8**(1), 43–63 (2012)
7. Chandrasekaran, S., Cooper, O., Deshpande, A., Franklin, M.J., Hellerstein, J.M., Hong, W., Krishnamurthy, S., Madden, S., Raman, V., Reiss, F., Shah, M.A.: Telegraphcq: Continuous dataflow processing for an uncertain world. In: Proceedings of the 1st Biennial Conference on Innovative Data Systems Research (CIDR-03) (2003)
8. Cugola, G., Margara, A.: Tesla: A formally defined event specification language. In: Proceedings of the 4th ACM International Conference on Distributed Event-Based Systems (DEBS-10), DEBS '10, pp. 50–61. ACM, New York, NY, USA (2010). DOI 10.1145/1827418.1827427
9. Della Valle, E., Ceri, S., Barbieri, D., Braga, D., Campi, A.: A first step towards stream reasoning. In: Proceedings of the 1st Future Internet Symposium (FIS 08), *LNCS*, vol. 5468, pp. 72–81. Springer (2009)
10. Dell'Aglio, D., Calbimonte, J., Valle, E.D., Oscar Corcho: Towards a unified language for RDF stream query processing. In: F. Gandon, C. Gueret, S. Villata, J.G. Breslin, C. Faron-Zucker, A. Zimmermann (eds.) The Semantic Web: ESWC-15 Satellite Events, Revised Selected Papers, *LNCS*, vol. 9341, pp. 353–363. Springer (2015). DOI 10.1007/978-3-319-25639-9_48
11. Dell'Aglio, D., Calbimonte, J.P., Balduini, M., Corcho, O., Della Valle, E.: On correctness in RDF stream processor benchmarking. In: H. Alani, L. Kagal, A. Fokoue, P. Groth, C. Biemann, J.X. Parreira, L. Aroyo, N. Noy, C. Welty, K. Janowicz (eds.) The Semantic Web – ISWC 2013: 12th International Semantic Web Conference, Sydney, NSW, Australia, October 21-25, 2013, Proceedings, Part II, pp. 326–342. Springer Berlin Heidelberg, Berlin, Heidelberg (2013). DOI 10.1007/978-3-642-41338-4_21
12. Dell'Aglio, D., Della Valle, E., Calbimonte, J., Corcho, O.: RSP-QL semantics: A unifying query model to explain heterogeneity of RDF stream processing systems. International Journal on Semantic Web and Information Systems (IJSWIS) **10**(4) (2015)
13. Galton, A.: Reified temporal theories and how to unreify them. In: Proceedings of the 12th International Joint Conference on Artificial Intelligence (IJCAI-91), vol. 2, pp. 1177–1182. Morgan Kaufmann Publishers Inc., San Francisco, CA, USA (1991)
14. Giese, M., Soylu, A., Vega-Gorgojo, G., Waaler, A., Haase, P., Jiménez-Ruiz, E., Lanti, D., Rezk, M., Xiao, G., Özçep, Ö.L., Rosati, R.: Optique: Zooming in on Big Data. IEEE Computer **48**(3), 60–67 (2015). DOI http://dx.doi.org/10.1109/MC.2015.82

15. Hwang, J.H., Xing, Y., Çetintemel, U., Zdonik, S.B.: A cooperative, self-configuring high-availability solution for stream processing. In: Proceedings of the 23rd IEEE International Conference on Data Engineering (ICDE 07), pp. 176–185 (2007)
16. Ioannidis, Y., Kotidis, Y., Mailis, T., Möller, R., Neuenstadt, C., Özçep, Ö.L., Svingos, C.: Deliverable D5.3 – data mining and query log analysis for scalable temporal and continuous query answering. Deliverable FP7-318338, EU (2015)
17. Jain, N., Mishra, S., Srinivasan, A., Gehrke, J., Widom, J., Balakrishnan, H., Çetintemel, U., Cherniack, M., Tibbetts, R., Zdonik, S.: Towards a streaming SQL standard. Proceedings of the VLDB Endowment **1**(2), 1379–1390 (2008)
18. Kharlamov, E., Brandt, S., Jiménez-Ruiz, E., Kotidis, Y., Lamparter, S., Mailis, T., Neuenstadt, C., Özçep, Ö.L., Pinkel, C., Svingos, C., Zheleznyakov, D., Horrocks, I., Ioannidis, Y.E., Möller, R.: Ontology-based integration of streaming and static relational data with Optique. In: F. Özcan, G. Koutrika, S. Madden (eds.) Proceedings of the 2016 International Conference on Management of Data (SIGMOD-16), pp. 2109–2112. ACM (2016). DOI 10.1145/2882903.2899385
19. Kharlamov, E., Kotidis, Y., Mailis, T., Neuenstadt, C., Nikolaou, C., Özçep, Ö.L., Svingos, C., Zheleznyakov, D., Ioannidis, Y., Lamparter, S., Möller, R., Waaler, A.: An ontology-mediated analytics-aware approach to support monitoring and diagnostics of static and streaming data. Journal of Web Semantics (2019). DOI https://doi.org/10.1016/j.websem.2019.01.001
20. Kharlamov, E., Kotidis, Y., Mailis, T., Neuenstadt, C., Nikolaou, C., Özçep, Ö.L., Svingos, C., Zheleznyakov, D., Brandt, S., Horrocks, I., Ioannidis, Y.E., Lamparter, S., Möller, R.: Towards analytics aware ontology based access to static and streaming data. In: P.T. Groth, E. Simperl, A.J.G. Gray, M. Sabou, M. Krötzsch, F. Lécué, F. Flöck, Y. Gil (eds.) Proceedings of the 15th International Semantic Web Conference (ISWC-16), Part II, *LNCS*, vol. 9982, pp. 344–362 (2016). DOI 10.1007/978-3-319-46547-0_31
21. Kharlamov, E., Mailis, T., Mehdi, G., Neuenstadt, C., Özçep, Ö.L., Roshchin, M., Solomakhina, N., Soylu, A., Svingos, C., Brandt, S., Giese, M., Ioannidis, Y., Lamparter, S., Möller, R., Kotidis, Y., Waaler, A.: Semantic access to streaming and static data at Siemens. Journal of Web Semantics pp. 54–74 (2017). DOI https://doi.org/10.1016/j.websem.2017.02.001
22. Kharlamov, E., Solomakhina, N., Özçep, Ö.L., Zheleznyakov, D., Hubauer, T., Lamparter, S., Roshchin, M., Soylu, A.: How semantic technologies can enhance data access at Siemens Energy. In: P. Mika, T. Tudorache, A. Bernstein, C. Welty, C. Knoblock, D. Vrandečić, P. Groth, N. Noy, K. Janowicz, C. Goble (eds.) Proceedings of the 13th International Semantic Web Conference (ISWC-14) (2014)
23. Krämer, J., Seeger, B.: Semantics and implementation of continuous sliding window queries over data streams. ACM Trans. Database Syst. **34**(1), 1–49 (2009). DOI 10.1145/1508857.1508861
24. Möller, R., Neuenstadt, C., Özçep, Özgür.L.: Deliverable D5.2 – OBDA with temporal and stream-oriented queries: Optimization techniques. Deliverable FP7-318338, EU (2014)
25. Möller, R., Neuenstadt, C., Özçep, Özgür.L.: Stream-temporal querying with ontologies. In: D. Nicklas, Özgür.L. Özçep (eds.) Proceedings of the First Workshop on High-Level Declarative Stream Processing (HiDeSt-15), co-located with KI-15, *CEUR Workshop Proceedings*, vol. 1447, pp. 42–55. CEUR-WS.org (2015)
26. Neuenstadt, C., Möller, R., Özçep, Özgür.L.: OBDA for temporal querying and streams with STARQL. In: D. Nicklas, Özgür.L. Özçep (eds.) Proceedings of the First Workshop on High-Level Declarative Stream Processing (HiDeSt-15), co-located with KI-15, *CEUR Workshop Proceedings*, vol. 1447, pp. 70–75. CEUR-WS.org (2015)
27. Özçep, Ö.L., Möller, R., Neuenstadt, C.: A stream-temporal query language for ontology based data access. In: Proceedings of the 37th German International Conference (KI-14), *LNCS*, vol. 8736, pp. 183–194. Springer International Publishing Switzerland (2014)
28. Özçep, Ö.L., Möller, R., Neuenstadt, C., Zheleznyakov, D., Kharlamov, E.: Deliverable D5.1 – a semantics for temporal and stream-based query answering in an OBDA context. Deliverable FP7-318338, EU (2013)

29. Özçep, Ö.L., Neuenstadt, C., Möller, R.: Deliverable D5.4—optimizations for temporal and continuous query answering and their quantitative evaluation. Deliverable FP7-318338, EU (2016)
30. Özçep, Ö.L., Möller, R.: Ontology based data access on temporal and streaming data. In: M. Koubarakis, G. Stamou, G. Stoilos, I. Horrocks, P. Kolaitis, G. Lausen, G. Weikum (eds.) Reasoning Web. Reasoning and the Web in the Big Data Era, *LNCS*, vol. 8714 (2014)
31. Özçep, Ö.L., Möller, R., Neuenstadt, C.: Stream-query compilation with ontologies. In: B. Pfahringer, J. Renz (eds.) Proceedings of the 28th Australasian Joint Conference on Artificial Intelligence (AI-15), *LNAI*, vol. 9457. Springer International Publishing (2015)
32. Phuoc, D.L., Dao-Tran, M., Parreira, J.X., Hauswirth, M.: A native and adaptive approach for unified processing of linked streams and linked data. In: Proceedings of the 10th International Conference on The Semantic Web (ISWC-11), Part I, pp. 370–388 (2011)
33. Soylu, A., Giese, M., Schlatte, R., Jiménez-Ruiz, E., Neuenstadt, C., Özçep, Ö.L., Brandt, S.: Querying industrial stream-temporal data: An ontology-based visual approach. Journal of Ambient Intelligence and Smart Environments **9**(1), 77–95 (2017)
34. Soylu, A., Giese, M., Schlatte, R., Jiménez-Ruiz, E., Özçep, Ö.L., Brandt, S.: Domain experts surfing on stream sensor data over ontologies. In: T.G. Stavropoulos, G. Meditskos, A. Bikakis (eds.) Proceedings of the 1st Workshop on Semantic Web Technologies for Mobile and Pervasive Environments, *CEUR Workshop Proceedings*, vol. 1588, pp. 11–20. CEUR-WS.org (2016)

Appendix

Proof of Theorem 6.6

In preparing the proof, I transform, for every $n \in \mathbb{N}$, the formulas in $\mathcal{L}^{\exists}_{HCL}$ into formulas (depending on n) that do not contain any quantifiers over the states of the sequence but may contain constants ι_j (for $j \in [n]$) denoting the states. Denote the rewriting by $\tau^n_{\iota_j}$ with parameters ι_j, n.

$\tau^n_{\iota_j}(\texttt{GRAPH i } \alpha) = \texttt{GRAPH j } \alpha$; $\quad \tau^n_{\iota_j}(\phi_1 \texttt{ AND } \phi_2) = \tau^n_{\iota_j}(\phi_1) \texttt{ AND } \tau^n_{\iota_j}(\phi_2)$
$\tau^n_{\iota_j}(\phi_1 \texttt{ OR } \phi_2) = \tau^n_{\iota_j}(\phi_1) \texttt{ OR } \tau^n_{\iota_j}(\phi_2)$; $\quad \tau^n_{\iota_j}(\phi_{arAt}) = \phi_{arAt}$
$\tau^n_{\iota_j}(\texttt{EXISTS i } \phi) = \tau^n_0(\phi) \texttt{ OR } \tau^n_1(\phi) \texttt{ OR } \dots \texttt{ OR } \tau^n_n(\phi)$
$\tau^n_{\iota_j}(\texttt{FORALL i } \phi) = \tau^n_0(\phi) \texttt{ AND } \tau^n_1(\phi) \texttt{ AND } \dots \texttt{ AND } \tau^n_n(\phi)$
$\tau^n(\phi) = \tau^n_{\iota_n}(\phi)$ (for any ϕ)

So the resulting formula $\tau^n(\phi)$ is equivalent to ϕ and it is made up by atoms of the form $\texttt{GRAPH } \iota_j\ \psi$ for ψ being a CQ and ι_j a constant denoting $j \in [n]$.

Without loss of generality I may assume that on either side of a conjunction the same set of variables occur. A semantic preserving transformation (adding identities) is possible. Let the resulting language be denoted $\mathcal{L}^{\exists,=}_{HCL}$. If $\mathcal{O} = \langle (\mathcal{A}_i)_{i \in [n]}, \mathcal{T} \rangle$ is not consistent, then we assumed that $cert_h(\phi, \mathcal{O}) = \text{NIL} = cert_{sep}(\phi, \mathcal{O})$. So in the following I may assume that $\mathcal{O}$ is consistent which means that for all $i \in [n]$ the local ontologies $\langle \mathcal{A}_i, \mathcal{T} \rangle$ are consistent.

I first verify that for all atoms ϕ that $cert_{sep}(\phi, \mathcal{O}) = cert_h(\phi, \mathcal{O})$ holds. This is trivial if ϕ is not a state atom. If ϕ is a state atom of the form $\texttt{GRAPH } \iota_i\ \alpha$ with a CQ α, then $cert_{sep}(\phi, \mathcal{O}) = cert_{sep}(\phi, \langle \mathcal{A}_i, \mathcal{T} \rangle) = cert(\phi, \langle \mathcal{A}_i, \mathcal{T} \rangle)$. Now

$cert(\phi, \langle \mathcal{A}_i, \mathcal{T} \rangle) = cert_h(\phi, O)$ holds due to the definitions of satisfaction relation in the case for state atoms and due to Proposition 6.4.

Now I show that $cert_h$ can be pushed through to the atoms by showing that it distributes over AND and OR. Let first $\phi = \phi_1$ AND ϕ_2 where the conjuncts have the same set of free variables, say $x_1, \ldots, x_n$. Then the following chain of equalities holds: $cert_h(\phi_1 \texttt{ AND } \phi_2, O) = \bigcap_{\hat{\mathfrak{I}} \models O} ans_{wh}(\phi_1 \texttt{ AND } \phi_2, \hat{\mathfrak{I}}) = \bigcap_{\hat{\mathfrak{I}} \models O} ans(\phi_1, \hat{\mathfrak{I}}) \cap \bigcap_{\hat{\mathfrak{I}} \models O} ans_{wh}(\phi_2, \hat{\mathfrak{I}})$. The latter is $cert_h(\phi_1, O) \cap cert_h(\phi_2, O)$. Now let $\phi = \phi_1$ OR ϕ_2 with free variables $x_1, \ldots, x_n$. It holds that

$$cert_h(\phi_1, O)) \cup cert_h(\phi_2, O)) \subseteq cert_h(\phi_1 \texttt{ OR } \phi_2, O)$$

For the other direction assume that $\mathbf{c} \in cert_h(\phi_1 \texttt{ OR } \phi_2, O)$; assume that $\mathbf{c} \notin cert_h(\phi_1, O) \cup cert_h(\phi_2, O)$. Then there must be models $\hat{\mathfrak{I}}_1, \hat{\mathfrak{I}}_2 \models O$ such that $\mathbf{c} \notin ans(\phi_1, \hat{\mathfrak{I}}_1)$ and $\mathbf{c} \notin ans(\phi_2, \hat{\mathfrak{I}}_2)$. It follows that $\hat{\mathfrak{I}}_1 \models \phi_2(\mathbf{c})$ and $\hat{\mathfrak{I}}_2 \models \phi_1(\mathbf{c})$. Now, $\mathbf{c}$ is also in the answer set for the canonical model $can(O)$, i.e., $\mathbf{c} \in ans(\phi_1 \texttt{ OR } \phi_2, can(O))$, so either $can(O) \models \phi_1(\mathbf{c})$ or $can(O) \models \phi_2(\mathbf{c})$. But as $can(O)$ is a universal model there is a homomorphism into every model of O, in particular for $\hat{\mathfrak{I}}_1$ and $\hat{\mathfrak{I}}_2$. As homomorphisms preserve positive existential clauses (see Proposition 2.1) one must have either $\mathbf{c} \in ans(\phi_1, \hat{\mathfrak{I}}_1)$ or $\mathbf{c} \in ans(\phi_2, \hat{\mathfrak{I}}_2)$, contradicting the assumption.

Chapter 7
Representation for Belief Revision

Abstract Belief revision deals with the problem of changing a declaratively specified repository under a new piece of potentially conflicting information. Ever since the first formal treatment of belief revision, representation theorems are considered important, not to say, canonical means to analyse belief revision operators. This chapter discusses representation theorems for a special kind of change operators, called reinterpretation operators, which resolve conflicts by disambiguating symbols. Moreover, it shows how various classical belief revision operators can be represented equivalently as reinterpretation operators.

7.1 Introduction

Belief revision [1] deals with the problem of changing a declaratively specified repository under a new piece of potentially conflicting information, called *trigger* or *trigger information* in the following. If there is trust in the incoming information—and classical belief revision in contrast to non-prioritised belief revision [16] assumes so—the integration may trigger a revision of the repository. The reason is that the trigger may be incompatible with the repository so that some of its formulae have to be eliminated in order to keep it satisfiable. An abstract means to describe the revision is by binary belief revision operators, i.e., functions getting a repository and a trigger as argument and outputting a new repository.

Depending on the intended role of the repository and the formal constraints on it, it is referred to under various names. If the repository is an arbitrary set of sentences (in some logic), then it is called a *belief base* [14]. If the repository is a logically closed set of sentences in some logic, then it is called a *belief set* [1]. If the main idea is to represent with the repository the knowledge of a domain by a finite belief base such that the revision operators applied on it depend only on the semantics and not the syntactical representation, then one talks of a *knowledge base* (for short KB) and of *knowledge base revision*—following the terminology of [6]. In this chapter I will mainly deal with operators for knowledge-base revision. In turn, within this class the

Ö. L. Özçep, *Representation Theorems in Computer Science*,
https://doi.org/10.1007/978-3-030-25785-9_7

focus is on knowledge-base revision operators that follow a particular construction principle, namely that of reinterpretation [9] as explained in the next sections.

In the last 30 years since the start of formal belief revision with the work of AGM (Alchourrón, Gärdenfors, and Makinson, [1]), roughly four construction principles were investigated and mutually interrelated. The first one is that of partial-meet revision going back to AGM [1]. The result of partial-meet revision is calculated by considering maximal subsets of the belief set consistent with the trigger and then intersecting a selection of them (hence the name partial-meet). Another construction principle (applied mainly to belief bases) uses kernels, i.e., minimal sets of inconsistencies, in order to guide the contraction/revision. The third principle rests on ranking sentences w.r.t. an epistemic entrenchment relation. And the last construction principle approaches revision purely semantically, considering only possible worlds or more specifically, models, for the KB and the trigger, taking into consideration relations (orderings) on the possible worlds to guide the revision. Knowledge-base revision falls into this last category.

For belief-revision operators relying on any of the construction principles above, the incompatibility between the KB and the trigger is explained with previously obtained false information in the KB. Therefore, the elimination of formulae in the knowledge base is an adequate means. In contrast, if the diagnosis for the incompatibility is not false information but ambiguous use of symbols, a different strategy and a corresponding construction principle seems more appropriate. As an example, consider an integration scenario where an ontology from an online library system (trigger) has to be integrated into an ontology of another online library system (the knowledge base). Here one faces the problem of ambiguous use of symbols. For example, the term "article" may be used in one library system for all entities that are published in the proceedings of a conference or in a journal, and in the other, "article" may stand for entities published in journals only. So one reading of "article" has to be reinterpreted such that conflicts due to ambiguity are resolved and such that the different readings are interrelated in a reasonable way. This idea of interrelating different readings is the core of semantic integration with semantic mappings, in particular: bridging axioms [23]. In the online library example, one would come up with a bridging axiom asserting that one reading of "article" is subsumed by the other reading.

The properties of reinterpretation heavily depend on the types of bridging axioms that one is going to allow to guide the reinterpretation. Choosing different classes of bridging axioms leads to different classes of reinterpretation operators. In the example above, allowing only equivalence (in propositional logic: bi-implication) may lead to a loss of relations between the different readings of "article" whereas a use of subsumption (in propositional logic: implication) may preserve relevant relations between the readings.

In this chapter I report on results for knowledge-base revision relying on reinterpretation—with a particular focus on representational aspects. For one class of revision operators based on reinterpretation, namely those using implications as bridging axioms, a representation theorem can be proved (see Sect. 7.4): there is a set of axioms—usually called *postulates* in the belief-revision community—that models exactly this class of reinterpretation operators. So, the represented class of structures are just the revision operators fulfilling the postulates and the representing class of structures is the same class, constructed with implications as bridging axioms.

Finding the right set of postulates is non-trivial due to the specific symbol- (and not sentence-) oriented construction principle for reinterpretation operators. When a symbol is reinterpreted, it is reinterpreted uniformly in the KB. Whereas classical belief revision eliminates only those sentences in the KB which are identified as culprits for an inconsistency, reinterpretation amounts to eliminating a whole bunch of formulae of the KB, namely those containing the reinterpreted symbol. In order to capture this uniformity condition, the main idea is to represent the KB by its most atomic components (formally: prime implicates)—thereby approximating the symbolic level by sentences as good as possible. Then, the effects of the implication-based operators are described by uniform closure conditions on the prime implicates.

As mentioned in the introductory chapter, it may be the case that there is not only one interesting construction principle for a set of axioms but more than one. And it is quite useful and instructive to consider different construction principles for the same operator because one may have additional or other properties not shared by the other system. For example, one of them might be more feasible than the other or might be better suited for implementation etc. In fact, regarding the point of implementability, this chapter gives an alternative representation for a class of operators based on reinterpretation. The construction principle for defining this class of operators requires the construction of possibly infinite sets of sentences, which makes a direct implementation impossible. But, as shown below, it is possible to describe the class of operators equivalently by finite operators that are more appropriate for implementation means (see Theorem 7.7): instead of referring to bridging axioms, the construction relies on flipping polarities of literals in the KB.

The last point regarding the multiplicity of different constructions is exploited even further in this chapter. Though the reinterpretation operators are mainly built to resolve ambiguities, such as those occurring in the integration scenarios illustrated above, one can argue that the reinterpretation-based framework is sufficiently general to capture also classical revision operators. This is an argument in favour of the fact that the hidden structure of reinterpretation is a useful one. I report on achieved results showing that it is possible to find appropriate classes of bridging axioms such that some of the classical revision operators discussed in [8] can be equally represented as reinterpretation operators: the revision operator of Weber [35] can be represented by full-meet revision on the set of bridging axioms that have the form of bi-implications. A natural variant of Weber revision can be represented as full-meet revision of bridging axioms having the form of implications. The operator of Satoh [34] (= skeptical operator of [7]) can be represented as partial-meet revision of the disjunctive closure of bi-implications, and the operator of Borgida [5] can be represented as partial-meet revision of bridging axioms that are either literals or contained in the disjunctive closure of bi-implications.

This chapter is structured as follows. Section 7.2 provides background on basic notions from belief revision. Section 7.3 discusses the general idea of reinterpretation and the class of reinterpretation operators investigated in this chapter. Section 7.4 gives a representation theorem for implication-based revision operators. Section 7.5 introduces the subclass of model-based revision operators for which the equivalence results with reinterpretation operators are proved in Sect. 7.6. The chapter is concluded by a discussion of related work (Sect. 7.7) and a summary (Sect. 7.8).

The content of this chapter is based on the papers [27, 30]. The framework of reinterpretation was developed in a series of papers [24, 25, 9, 29, 28] in the context of my PhD thesis [26]. But neither in the series of papers nor in the thesis I considered representational aspects in detail, which, in turn, are at the core of this monograph.

7.2 Preliminaries

For the results and the construction of reinterpretation operators investigated in this chapter I draw on various constructions and results from classical revision theory as initiated by Alchourrón, Gärdenfors, and Makinson (AGM) [1]. The revision operators of AGM rest on a notion of logic in the polish tradition. A logic is represented as a binary tuple $(\mathcal{L}, Cn(\cdot))$ with a set of formulae $\mathcal{L}$ and a consequence operator $Cn(\cdot)$. (Remember from Chap. 2 that a consequence operator is reflexive, monotone, and idempotent.) Further, AGM assumes that $\mathcal{L}$ contains all propositional formulae and additionally requires the consequence operator to be supra-classical, i.e., if α follows classically from X, then $\alpha \in Cn(X)$, to fulfil the deduction theorem, i.e., if some formula β follows from X and α, then $\alpha \rightarrow \beta$ follows from X, and to be compact, i.e., if α follows from X, then it already follows from a finite subsets of X.

In all of the following considerations of this chapter the set of formulae $\mathcal{L}$ is the set of sentences $sent(\mathcal{P})$ for a set of propositional symbols $\mathcal{P}$. And the consequence operator $Cn(\cdot)$ is the one induced by the entailment relation $\models$ for propositional logic. Actually I will switch between the use of the consequence operator and the use of the entailment relation, using the fact that $\alpha \in Cn(X)$ iff $X \models \alpha$.

Next to this consequence operator I consider another operator that is a weakening of classical propositional consequence: some of the equivalence results in this chapter heavily depend on the use of the disjunctive closure [17]. The *disjunctive closure* $\overline{B}$ of a finite set of sentences B is defined as follows:

$$\overline{B} = \{\beta_1 \vee \cdots \vee \beta_n \mid \beta_i \in B, n \in \mathbb{N} \setminus \{0\}\}$$

Classical belief-revision functions à la AGM operate on belief sets, i.e., sets logically closed w.r.t. the consequence operator, and a formula which triggers the revision of the belief set into a new belief set. Some of the reinterpretation operators are based on revising logically closed sets of bridging axioms. But note the difference: in reinterpretation not the repository (belief-set, belief base or knowledge base) is revised, but a closure of bridging axioms.

Reinterpretation operators are defined with *dual remainder sets*, which are similar to the concept of remainder sets by AGM [1]. The notion of a dual remainder set is a generalisation that can be used to define revision operators also for logics that do not allow for sentential negation. (See [24, 25, 9, 29, 28] for examples of applying the idea of reinterpretation for revising KBs in description logic, where negation is a concept constructor and not a sentence constructor.)

As I am going to consider also the more general case of *multiple revision*, i.e., revision with sets Y of sentences as triggers, I define the following notions for this general case. The special case of singletons $Y = \{\alpha\}$ covers the case of a trigger that is a sentence α. *The dual remainder sets modulo* Y are defined as follows:

$$X \in B \top Y \text{ iff } X \subseteq B \text{ and } X \cup Y \text{ is consistent and}$$
$$\text{for all } X' \subseteq B \text{ with } X \subsetneq X' \text{ the set } X' \cup Y \text{ is not consistent}$$

Let B be an arbitrary set of sentences. An *AGM-selection function* γ for B is a function $\gamma : pow(B) \longrightarrow pow(B)$, such that for all sets of formulae Y the following holds: if $B \top Y \neq \emptyset$, then $\emptyset \neq \gamma(B \top Y) \subseteq B \top Y$, else $\gamma(\emptyset) = \{B\}$.

With these notions one can define multiple partial-meet revision for arbitrary sets of sentences B, i.e., belief bases (see, e.g., [17]). Given any set B, an AGM-selection function γ for B, and any set Y of formulae, *multiple partial-meet base revision* is defined as:

$$B *_\gamma Y = \bigcap \gamma(B \top Y) \cup Y$$

Revising with a single sentence α then is nothing else than revising with the singleton $\{\alpha\}$, i.e.: $B *_\gamma \alpha = B *_\gamma \{\alpha\}$. Two special cases of partial-meet revision result from two instantiations of the selections function γ: If γ selects all of the sets, then the revision is called *full-meet revision* and has the form $B * Y = B *_\gamma Y = \bigcap B \top Y \cup Y$. If γ selects exactly one set, then the revision is called *maxi-choice revision*.

For the representation theorem of Sect. 7.4 I am going to set up a set of postulates. These are motivated by postulates discussed for belief-base revision. As the focus is on knowledge-base revision (where the syntactical representation of the KB is not allowed to influence the revision) these postulates will be adapted in quite a strong way. The adaptations will be discussed with reference to the following postulates of classical belief-base revision.

(BR1) $B * \alpha \not\models \bot$ if $\alpha \not\models \bot$.
(BR2) $\alpha \in B * \alpha$.
(BR3) $B * \alpha \subseteq B \cup \{\alpha\}$.
(BR4) For all $\beta \in B$ either $B * \alpha \models \beta$ or $B * \alpha \models \neg\beta$.
(BR5) If for all $B_1 \subseteq B$: $B_1 \cup \{\alpha\} \models \bot$ iff $B_1 \cup \{\beta\} \models \bot$, then $(B * \alpha) \cap B = (B * \beta) \cap B$.

Postulate (BR1) is the *consistency postulate* [1]. It says that the revision result has to be consistent in case the trigger α is consistent. Postulate (BR2) is the *success postulate* ([12]): the revision must be successful in so far as α has to be in the revision result. (BR3) is called the *inclusion postulate* for belief-base revision [17, p. 200]. The revision result of operators fulfilling it are bounded from above. Postulate (BR4) is the *tenacity postulate* [13]. It states that the revision result is complete with respect to all formulae of B [13]. Postulate (BR5) is the *logical uniformity postulate* for belief-base operators [15]. It says that the revision outcomes are determined by the subsets (in)consistent with the trigger. The logical uniformity postulate generalises the right extensionality postulate for revision operators, which states that the revision outcomes of equivalent triggers α, β lead to the same revision result: $B * \alpha = B * \beta$ if $\alpha \equiv \beta$. In contrast to Hansson, I specified the postulate as "logical" as I will use the notion of uniformity later on in a different sense.

Reinterpretation operators are based on bridging axioms. Two classes of bridging axioms used throughout this chapter are the set of bi-implications

$$Bimpl = \{\overleftrightarrow{p} \mid p \in \mathcal{P}\}$$

and the set of implications

$$Impl = \{\overrightarrow{p}, \overleftarrow{p} \mid p \in \mathcal{P}\}$$

As I am going to define quite a lot of change operators, I make the following notational convention: All revision operators are denoted by $*$, possibly with super- and subscripts. All reinterpretation operators are denoted by the symbol $\circ$, possibly with super- and subscripts.

7.3 Reinterpretation Operators

The general idea of applying a reinterpretation operator on a knowledge base B and a trigger α is to trace back the conflict between B and the trigger α to an ambiguous use of some of the common symbols. So, the idea for resolving the conflict is to assume in the first place how the different uses of the symbols are related, stipulating the relations explicitly as a set of bridging axioms BA, and then applying a classical revision strategy on BA as the knowledge base to be revised.

Example 7.1 Assume that the sender of the trigger has a strong notion of article, namely that of objects published in a journal. The sender of the trigger has stored the information that some entity b is not an article in her KB because she has acquired the knowledge that b is not published in a journal. Assume that the information $\neg Article(b)$ is represented in propositional logic by the literal $\alpha = \neg q$. This bit of information α is sent to the holder of the KB B. The holder of B has a weaker notion of article—defining them as entities published either in the proceedings of a conference or in a journal. In her KB the same entity b is stated to be an article. So B entails q. Now, a conflict resolution strategy is to completely separate the readings of all symbols by renaming all the ones in B with primed versions, resulting in an "internalised" KB B'. In particular, q becomes q' in the receiver's KB. Then, guesses on the interrelations of the different readings are postulated by bridging axioms. The type of reinterpretation depends on the class of initial bridging axioms. So for example, considering bi-implications would lead to stipulations of axioms $p \leftrightarrow p'$, actually stating that the reading of p and p' is the same. The resolution of the conflicts between B and α requires not to include $q \leftrightarrow q'$, as this entails an inconsistency.

Considering more fine-grained sets of bridging axioms such as implications $p \rightarrow p'$ and $p' \rightarrow p$ leads to more fine-grained solutions. In this example, the reinterpretation result could contain $q \rightarrow q'$ (articles as used in the trigger are articles as used in B) but not $q' \rightarrow q$ (articles in the sense of B are articles in the sense of α).

The following parameterised equation illustrates the general strategy for reinterpretation:

$$B \circ \alpha = BA * (B' \cup \{\alpha\})$$

The first parameter is the set of initial bridging axioms: BA contains sentences over $\mathcal{P} \cup \mathcal{P}'$ relating the meanings of the symbols $\mathcal{P}$ (associated with the sender) with those in $\mathcal{P}'$ (associated with the receiver). A very simple example of a bridging axiom is the bi-implication $p \leftrightarrow p'$ stating that the reading of p in the knowledge base is actually the same as the reading in the trigger. The trigger is the union of the original trigger and the "internalised version" B' of the original knowledge base B. Concretely, here and in the following B' is the outcome of substituting in every sentence of B propositional symbols p with their primed variants p'. The second parameter in the schema is a classical base-revision function $*$. In this chapter, only partial-meet revision on arbitrary finite KBs is considered as instance of $*$.

Technically, reinterpretation-based revision is similar to base-generated revision [17] which combines the benefits of belief-base revision and belief-set revision, namely: the benefit of having a finite (and hence implementable) resource and the benefit of syntax-insensitivity. The difference relies on the special type of the generating base one uses for reinterpretation, namely a set of bridging axioms. As the reinterpretation-based approach is not sentence-oriented but symbol-oriented it can be applied to various logics such as description logics [9].

Four different classes of reinterpretation operators fitting the above schema are those based on bi-implications as bridging axioms and those based on implications, both in turn considered per se or w.r.t. the disjunctive closure.

Definition 7.2 Let B be a knowledge base, α be a formula, and γ be a selection function for *Impl* ($\overline{Impl}$, *Bimpl*, $\overline{Bimpl}$, resp.) and $*_\gamma$ be a partial-meet revision operator for *Impl* ($\overline{Impl}$, *Bimpl*, $\overline{Bimpl}$, resp.). 1. The *implication-based*, 2. the *disjunctively closed implication-based*, 3. the *bi-implication based*, and 4. the *disjunctively closed bi-implication based reinterpretation operators* are defined as follows

1. $B \circ_\gamma^{\rightarrow} \alpha = Impl *_\gamma (B' \cup \{\alpha\})$
2. $B \circ_{\underline{\gamma}}^{\rightarrow} \alpha = \overline{Impl} *_\gamma (B' \cup \{\alpha\})$
3. $B \circ_\gamma^{\leftrightarrow} \alpha = Bimpl *_\gamma (B' \cup \{\alpha\})$
4. $B \circ_{\underline{\gamma}}^{\leftrightarrow} \alpha = \overline{Bimpl} *_\gamma (B' \cup \{\alpha\})$

If γ is the identity function, then γ can be dropped and the resulting operators are called *skeptical reinterpretation operators* (using the terminology of [7]). If γ is such that $|\gamma(X)| = 1$ for all X, then the induced operator is called a *choice reinterpretation* operator.

Another reinterpretation operator—which does not fit into the homogeneous scheme of Definition 7.2 and hence is defined separately—is termed *literal-supported* reinterpretation operator. It uses the notion of a bridging axiom in a very tolerant way. Concretely, the operator uses the following set of bridging axioms:

$$Bimpl^{+} = \overline{Bimpl} \cup \overline{\{p', \neg p' \mid p \in \mathcal{P}\}}$$

So, next to the disjunctive closure of bi-implications it contains the disjunctive closure of literals in the internal vocabulary.

Definition 7.3 Let γ be an AGM-selection function for $Bimpl^{+}$ and $*_{\gamma}$ a multiple partial-meet revision operator for $Bimpl^{+}$. Then the *literal-supported* reinterpretation operators are defined by: $B \circ_{\gamma}^{lit} \alpha = Bimpl^{+} *_{\gamma} (B' \cup \{\alpha\})$.

Note that the result of all reinterpretation operators introduced above contain bridging axioms and hence are not subsets of $sent(\mathcal{P})$ but subsets of $sent(\mathcal{P} \cup \mathcal{P}')$. Accordingly, reinterpretation operators are not genuine revision operators. This is prima facie not an essential problem because for any reinterpretation operator $\circ$ a corresponding revision $*_{\circ}$ operator is definable by restricting the outcome to the vocabulary $\mathcal{P}$ of the sender. But there are different ways to implement the restriction: One extreme case is to delete from the result all sentences that contain at least one internal symbol (i.e. a symbol from $\mathcal{P}'$). Technically, this amounts to the definition $B *_{\circ} \alpha = (B \circ \alpha) \cap sent(\mathcal{P})$. On the other extreme is the solution that, first, one closes the resulting set w.r.t. the consequence operator and only afterwards one does the restriction, technically: $B *_{\circ} \alpha = Cn^{\mathcal{P}}(B \circ \alpha) = Cn^{\mathcal{P} \cup \mathcal{P}'}(B \circ \alpha) \cap sent(\mathcal{P})$. And of course, there are even further possibilities when considering other closure operators.

Because of these possibilities, I decided to stick to the distinction between reinterpretation operators and revision operators and just compare them w.r.t. the consequences in the public vocabulary $\mathcal{P}$. Equivalence results w.r.t. the consequences in the public vocabulary are the contents of Sect. 7.6. In that section I show that reinterpretation operators are sufficiently expressive in order to capture classical revision operators. This means actually that for a given revision operators $*$ one can find a reinterpretation operator $\circ_{*}$ such that for all B and α the equivalence $B * \alpha \equiv_{\mathcal{P}} B \circ_{*} \alpha$ holds. This equivalence can equally be expressed as $[\![B * \alpha]\!] = [\![Cn^{\mathcal{P}}(B \circ_{*} \alpha)]\!]$.

All reinterpretation operators considered in this chapter are motivated by a specific integration scenario such as the one discussed in the example above on online library systems: The information in the KB and that in the trigger are over the same domain. There is trust in the information stemming from the sender of the trigger and there is a clear evidence that the symbols in the KB and in the trigger are strongly related—though they still may differ. This is the reason why, in this chapter, only special kinds of bridging axioms are considered where one reading p is related to another syntactically similar reading p'. Clearly, one can consider further bridging axioms that go beyond the resolution of ambiguities. For example, one may consider also bridging axioms that relate synonymous symbols (say "beverage" and "drink"). But nothing prevents the general reinterpretation framework from using these kinds of bridging axioms. Of course, what is required then is a knowledge engineering step (based on heuristics, say) regarding the potential conflicts in a given integration scenario in order to find an appropriate initial set of bridging axioms.

The idea of reinterpretation, as developed in [26] was not completely new. It was used implicitly in the operators of Delgrande and Schaub (DS) [7], but not from the perspective of disambiguating symbols, rather from the perspective of using bridging axioms as auxiliary axioms for revision. Moreover, though the conflict resolution in DS revision is similar to reinterpretation it is not the same. Due to the implicit use of reinterpretation, I give the definitions of the revision operators of DS already in this section. Other relevant revision operators are defined in Sect. 7.5

DS revision operators are defined with the notion of a *belief-change extension.* I describe here only their general framework for revision and not that of parallel revision and contraction in a belief change scenario.

Definition 7.4 ([7]) Given B and α, a *belief-change extension*, for short: a bc extension, is defined as a set of the form $Cn^{\mathcal{P}}(B' \cup \{\alpha\} \cup Bim_i)$ where $Bim_i \in Bimpl\top(B' \cup \{\alpha\})$. If no such Bim_i exists, then $sent(\mathcal{P})$ is the only belief-change extension. The family of all belief-change extensions is denoted by $(E_i)_{i \in I}$.

A DS-selection function c is defined for I as $c(i) \in I$. So it corresponds to AGM-selection functions that select exactly one element.

Based on these notions, choice revision $*^c_{\mathrm{DS}}$, which selects exactly one bc extension, and skeptical revision, which selects all bc extensions, can be defined.

Definition 7.5 ([7]) Given a KB B, a formula α, the set of all bc extensions $(E_i)_{i \in I}$, and a selection function over I with $c(I) = k$, *choice revision* $*^c_{\mathrm{DS}}$ and skeptical revision $*_{\mathrm{DS}}$ are defined as follows:

$$B *^c_{\mathrm{DS}} \alpha = E_k (\text{for } c(I) = k) \text{ and } B *_{\mathrm{DS}} \alpha = \bigcap_{i \in I} E_i$$

The idea of using bc extensions can also be used for other sets than bi-implications. This may result in the following definitions as described in [27]. Let us call a set $Cn^{\mathcal{P}}(B' \cup \{\alpha\} \cup X)$ an *implication-based bc extension* iff $X \in Impl\top(B' \cup \{\alpha\})$. Let $(Im_i)_{i \in I}$ be the set of all implication-based consistent bc extensions for B and α, and let c be a selection function for I with $c(I) = k$. The new operators are defined as follows:

Definition 7.6 *Implication-based choice revision* $*^{c,\rightarrow}_{\mathrm{DS}}$ and *implication-based skeptical revision* $*^{\rightarrow}_{\mathrm{DS}}$ are defined by:

$$B *^{c,\rightarrow}_{\mathrm{DS}} \alpha = Im_k \text{ (for } c(I) = k) \text{ and } B *^{\rightarrow}_{\mathrm{DS}} \alpha = \bigcap_{i \in I} Im_i$$

The implication-based choice-revision operator $*^{c,\rightarrow}_{\mathrm{DS}}$ is the one which is in the main focus of this chapter. From a representational view it can be characterised in three ways: It can be characterised by a set of postulates as shown in Sect. 7.4. It can equivalently be characterised by an implication-based choice reinterpretation operator $\circ_{\gamma}^{\rightarrow}$ as shown in Proposition 7.22. A last form of representation is given in the following theorem and concerns the fact that the result of implication-based choice revision is an infinite set. It can be shown that the infinity of the result is a mere technicality in the sense that there is an operator that gives a finite result and is equivalent to implication-based choice revision. In order to state and prove this result, I have to define the relevant notion of partial flipping here, which is an adaptation of the flipping operator used in [7] for proving a finite representation result for bi-implication based revision.

For ease of definition I assume that only connectors $\wedge, \vee$ and $\neg$ appear in the knowledge base (otherwise one just has to recompile the KB equivalently). An occurrence of a propositional symbol is *syntactically positive* iff it occurs in the scope of an even number of negation symbols, otherwise it is *syntactically negative*. Let $(Im_i)_{i \in I}$ be the family of bc extensions for B and α, and let Im_k be an implication-based bc extension chosen by the selection function, $c(I) = k$. The result of *partial flipping* over B, for short $\lceil B \rceil_k^{\rightarrow}$, is defined as follows: If $p \rightarrow p' \notin Im_k$, then switch the polarity of the negative occurrences of p in $\bigwedge B$ (by adding $\neg$ in front of these occurrences). If $p' \rightarrow p \notin Im_k$, then switch the polarity of the positive occurrences of p in $\bigwedge B$. Let $\lceil B \rceil^{\rightarrow} = \bigvee_{i \in I} \lceil B \rceil_i^{\rightarrow}$.

With these definitions one can state the finite representability theorem.

Theorem 7.7 ([27]) *$B *_{DS}^{c,\rightarrow} \alpha$ has the same models as $\lceil B \rceil_c^{\rightarrow} \wedge \alpha$ and $B *_{DS}^{\rightarrow} \alpha$ has the same models as $\lceil B \rceil^{\rightarrow} \wedge \alpha$.*

Proof See p. 173. □

7.4 A Representation Theorem for Implication-Based Choice Revision

This section develops the theory that is necessary to yield a representation theorem for implication-based choice-revision operators $*_{\mathrm{DS}}^{\rightarrow}$.

The main distinctive feature of Delgrande's and Schaub's operators $*_{\mathrm{DS}}^{c}$, $*_{\mathrm{DS}}$ as well as of $*_{\mathrm{DS}}^{c,\rightarrow}$, $*_{\mathrm{DS}}^{\rightarrow}$ is that these are defined on finite sets B of formulae as left argument, but do not depend on the specific representation of B, i.e., they are knowledge-base revision operators [6]. My strategy is to adapt known postulates for belief-base revision (see preliminaries of this chapter) to knowledge-base revision. For this, one has to replace all references to the set B and its subsets by syntax-insensitive concepts.

The key for the adaptation is the use of prime implicates entailed by the knowledge base B. Roughly, prime implicates are the most atomic clauses entailed by B. In the following subsection I recapitulate the definition of prime implicates $prime(B)$ for a knowledge base B and restate the fact that B is equivalent to $prime(B)$ (Proposition 7.8).

A second adaptation concerns the uniformity of the operators $*^{c}_{\text{DS}}$, $*_{\text{DS}}$ as well of the operators $*^{c,\rightarrow}_{\text{DS}}$, $*^{\rightarrow}_{\text{DS}}$. The conflicts between B and the trigger α are handled on the level of symbols and not on the level of formulae. Therefore, in order to mirror this effect on the prime implicates one has to impose a uniformity condition. This will be done implicitly by switching the perspective even further from prime implicates to uniform sets of prime implicates (see Definition 7.10 below).

7.4.1 Prime Implicates and Uniform Sets

Let $\mathcal{P}$ be a set of propositional symbols and $S \subseteq \mathcal{P}$. Let $\alpha \in sent(\mathcal{P})$. Let α be a non-tautological formula. The set $prime^S(\alpha)$ of prime implicates of α over S is defined in the following way.

$$prime^S(\alpha) = \{\beta \in clause^S(\alpha) \mid \emptyset \not\models \beta \text{ and } \beta \text{ has no proper subclause in } clause^S(\alpha)\}$$

So, the set of prime implicates of α w.r.t. S is the set of all minimal clauses in $sent(S)$ following from α. For tautological formulae α let $prime^S(\alpha) = \{p \vee \neg p\}$, where p is the first propositional symbol occurring in α with respect to a fixed order of $\mathcal{P}$. For knowledge bases let $prime^S(B) = prime^S(\bigwedge B)$. If S is clear from the context, then I just write $prime(\alpha)$. The conjunction of all formulae in $prime(\alpha)$ is called the *Dual Blake Canonical Form (DBCF)* of α [2].[1]

Clearly, the set of prime implicates of a knowledge base B is equivalent to B itself. In the following propositions I use the abbreviation $prime(\cdot) = prime^{\mathcal{P}}(\cdot)$ and assume that $B, B_1, B_2 \subseteq sent(\mathcal{P})$.

Proposition 7.8 *For knowledge bases B:* $prime(B) \equiv B$.

Proof See p. 174. □

An additional simple fact is given in the following proposition. It justifies the perspective on the set of prime implicates as a canonical representation for the knowledge contained in the knowledge base.

Proposition 7.9 *If* $B_1 \equiv B_2$, *then* $prime(B_1) = prime(B_2)$.

Proof See p. 174. □

The notion of uniform sets is introduced in order to capture the conflict resolution strategy by the implication based revision operators. If, e.g., it holds that the bridging axiom $p' \rightarrow p$ is eliminated in the conflict resolution process, then formulae of the knowledge base B, in which p occurs positively, are not preserved in the revision result. In general, if a set of implication-based bridging axioms Im is given, then

[1] The definition of prime implicates according to [2] does not explicitly exclude tautologies. But their examples do not contain tautologies. Therefore I excluded tautological clauses, too.

$B' \cup Im$ preserves a subset of prime implicates of B which fulfils some closure condition concerning the polarities of symbols. These sets of prime implicates can be characterise as uniform sets according to the following definition.

Definition 7.10 Let $B \subseteq sent(\mathcal{P})$ be a knowledge base. A set $X \subseteq prime(B)$ is called *uniform w.r.t. B and implications*, $X \in U^{Impl}(B)$ for short, iff the following closure condition holds: if $pr \in prime(B)$ is such that (a) $symb(pr) \subseteq symb(X)$ and (b) for all symbols p in pr there is a $pr_p \in X$ that contains p in the same polarity, then pr is contained in X, i.e., $pr \in X$.

Example 7.11 Let $B = \{p \vee q, p \vee r \vee s, r \vee t, s \vee u\}$. Then $prime(B) = B$. Now, among all subsets $X \subseteq prime(B)$ only the set $Y := \{p \vee q, r \vee t, s \vee u\}$ is not uniform as it would have to contain $p \vee r \vee s$, too. Formally, $U^{Impl}(B) = pow(prime(B)) \setminus \{\{p \vee q, r \vee t, s \vee u\}\}$. Note that only in case of the non-uniform set Y one cannot find a set of implication based hypotheses Im such that Y is exactly the set of prime implicates of B which are preserved by $B' \cup Im$.

The following proposition on the closure of the set of uniform sets is an immediate consequence of their definition.

Proposition 7.12 *For all $X, Y \in U^{Impl}(B)$ it is the case that $X \cap Y \in U^{Impl}(B)$.*

The union of uniform sets may not be uniform again. So let $X \cup' Y$ denote the smallest uniform set containing the uniform sets X and Y.

The interaction of uniform sets with implications, which will be used for the representation theorem below, is stated in the following propositions. The first proposition states how prime implicates interact with uniform substitutions. Uniform substitutions ρ are (partial) functions from propositional symbols to propositional formulae. Expression $\rho(\alpha)$ or $\alpha\rho$ results from α by substituting for all propositional symbols p all its occurrences in α by the formula $\rho(p)$.)

Proposition 7.13 *Let $\mathcal{P}$ and $\mathcal{P}'$ be disjoint sets of propositional symbols. Let B be a knowledge base and ρ be a uniform injective substitution for some subset $S = \{p_1, \ldots, p_n\} \subseteq \mathcal{P}$ such that $\rho(S) = \{p'_1, \ldots, p'_n\} \subseteq \mathcal{P}'$.*
Then: $prime^{\mathcal{P} \cup \mathcal{P}'}(B\rho) \equiv_{\mathcal{P}} prime^{\mathcal{P}}(B\rho)$.

Proof See p. 174. □

The interaction of a set of implications Im with prime implicates is captured in the following proposition. It states that all prime implicates of $B' \cup Im$ that do not contain primed symbols are prime implicates of B.

Proposition 7.14 *Let $\mathcal{P}$ and $\mathcal{P}'$ be disjoint sets of propositional symbols. Let B be a knowledge base and ρ be a uniform injective substitution for a set $S = \{p_1, \ldots, p_n\} \subseteq \mathcal{P}$ such that $\rho(S) = \{p'_1, \ldots, p'_n\} \subseteq \mathcal{P}$ and let Im be a set of implication based hypotheses containing at most primed symbols of $\rho(S)$.*
Then: $prime^{\mathcal{P}}(B\rho \cup Im) \subseteq prime^{\mathcal{P}}(B)$.

Proof See p. 175. □

As a corollary to the propositions, one can deduce the main result of this subsection, Theorem 7.15. This theorem is a proper justification for Definition 7.10—in the sense that Definition 7.10 really captures the intended concept. The theorem shows that for all B and Im one can find a uniform set X that is equivalent to $B' \cup Im$. The set X exactly describes the collection of logical atoms (prime implicates) of the receiver's KB B that are preserved after dissociating the name spaces of the sender and receiver (step from B to B') and adding hypotheses on the semantical relatedness in Im.

Theorem 7.15 *Let $\mathcal{P}$ and $\mathcal{P}'$ be disjoint sets of propositional symbols. Let B be a knowledge base and ρ be a uniform injective substitution for some subset $S = \{p_1, \ldots, p_n\} \subseteq \mathcal{P}$ such that $\rho(S) = \{p'_1, \ldots, p'_n\} \subseteq \mathcal{P}$ and let Im be a set of implication based hypotheses containing at most primed symbols of $\rho(S)$.*
Then there is a uniform set $X \in U^{Impl}(B)$ such that: $B' \cup Im \equiv_{\mathcal{P}} X$.

Proof See p. 177. □

7.4.2 Postulates for Implication-Based Choice Revision

The following postulates for revision operators $*$ are adapted variants of the postulates mentioned in this chapter's section on preliminaries (Sect. 7.2). They are exactly the ones that characterise the implication-based choice revision operators.

(R1) $B * \alpha \not\models \bot$ if $B \not\models \bot$ and $\alpha \not\models \bot$.
(R2) $B * \alpha \models \alpha$.
(R3) There is a set $H \subseteq U^{Impl}(B)$ such that $B*\alpha \equiv \bigwedge \bigcup' H \wedge \alpha$ or $B*\alpha \equiv \bigwedge \bigcup' H$.
(R4) For all $X \in U^{Impl}(B)$ either $B * \alpha \models X$ or $B * \alpha \models \neg \bigwedge X$.
(R5) For all $Y \subseteq U^{Impl}(B)$: if $\bigcup' Y \cup \{\alpha\} \models \bot$ iff $\bigcup' Y \cup \{\beta\} \models \bot$, then $\{X \in U^{Impl}(B) \mid B * \alpha \models X\} = \{X \in U^{Impl}(B) \mid B * \beta \models X\}$.

Postulate (R1) can be termed the postulate of weak consistency. It says that the revision result has to be consistent (satisfiable) in case both the trigger α and the knowledge base B are consistent. The consistency postulate for belief-base revision (BR1) is stronger as it requires consistency also in the case where only α is consistent. Postulate (R2) is a weak success postulate: the revision must be successful in so far as the result has to entail (and not necessarily contain) α. (R3) is an adapted version of the inclusion postulate for belief-base revision (BR3). The classical inclusion postulate can be rewritten as: there is a $B_1 \subseteq B$ such that $B * \alpha = B_1 \cup \{\alpha\}$ or $B * \alpha = B_1$. In (R3) B is replaced by the set of uniform sets w.r.t. B, and set identity is shifted to equivalence. Note that due to the definition of the $\cup'$-closure operator the set $\bigcup' H'$ is a uniform set and hence the postulate (R3) can be reformulated as:

(R3') There is a set $H \in U^{Impl}(B)$ such that $B * \alpha \equiv \bigwedge H \wedge \alpha$ or $B * \alpha \equiv \bigwedge H$.

Postulate (R4) can be called *uniform tenacity*. It is a very strong postulate, which states that all uniform sets w.r.t. B either follow from the result or are falsified.

This postulate captures the maximality of the operator $*_{\mathrm{DS}}^{c,\rightarrow}$. Postulate (R5) is an adaptation of the logical uniformity postulate for belief-base operators (BR5). It says that the revision outcomes w.r.t. the revision operator $*$ are determined by the uniform sets entailed by the revision result.

As the representation theorem below shows, postulates (R1)–(R5) are sufficient to represent the class of implication-based choice revision operators modulo equivalence.

Theorem 7.16 *A revision operator $*$ fulfils the postulates (R1)–(R5) iff it can be equivalently described as $*_{DS}^{c,\rightarrow}$ for some selection function c.*

Proof See p. 177. □

7.5 Model-Based Belief Revision

The purpose of this and the next section is to demonstrate that some well-known classical belief-revision operators can be simulated by reinterpretation operators. In fact, all of the revision operators, which I represent equivalently as reinterpretation operators in the next section, can be described as model-based belief revision operators.

The main idea of *model-based belief revision* is to drive the revision only by the models of the knowledge base and of the trigger—and nothing else. With this approach the concrete syntactical representation of the belief base becomes irrelevant and hence model-based revision can be considered as a proper means to implement knowledge-base revision operators: given a KB B, consider its models $[\![B]\!]$, apply a model-based revision operator on it, and then finitely represent the outcome again as a new KB.

Though quite many different model-based operators exist, the core idea for the revision is the same: the models of the revision are those models of the trigger that are minimal w.r.t. some appropriate (pre-, partial, or total) order or, more specifically, a distance function. Formally, model-based revision can be described by the following equation:

$$B * \alpha = FinRep(Min_{\leq_{[\![B]\!]}}([\![\alpha]\!]))$$

Figure 7.1 illustrates the general idea of model-based revision. The small dots and rectangles stand for propositional assignments (or more generally for possible worlds). The small rectangles are those models of α with minimal distance from the set of models of the KB B. The dotted circles indicate different distances from the models of the belief base B.

As shown by Katsuno and Mendelzon [18], there are strong connections between revision operators based on orders (of a specific kind) and the postulates they fulfil.

All model-based operators that are in the focus of this and the following sections are defined on the base of minimal difference between models of the knowledge base and the trigger. Minimal difference in turn is explicated by using—in some or

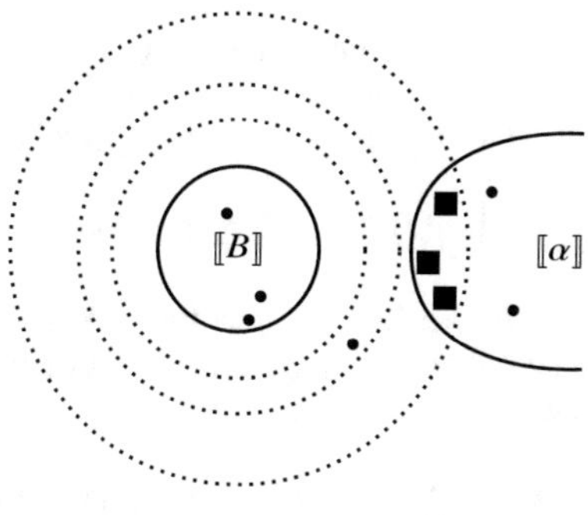

Fig. 7.1 Illustration of model-based revision

other form—the symmetric difference of models represented as sets. The *symmetric difference* for any pair of sets A, B is defined as $A \Delta B = A \setminus B \cup B \setminus A$. Let X_1, X_2 be sets of sentences. $\Delta^{min}(X_1, X_2)$ is the set of inclusion-minimal symmetric differences between models of X_1 and X_2.

$$\Delta^{min}(X_1, X_2) = min_{\subseteq}\{\Im \Delta \Im \mid \Im \in [\![X_1]\!], \Im \in [\![X_2]\!]\}$$

If one considers a set of sentences X_1 with exactly one model $\Im$ in the first argument, then one gets as special case $\Delta^{min}(\Im, X_2)$ which is the set of inclusion-minimal symmetric differences between $\Im$ and models of X_2. (Remember that $\Im$ is identified with the set of proposition symbols that are true according to $\Im$).

$$\Delta^{min}(\Im, X_2) = min_{\subseteq}\{\Im \Delta \Im \mid \Im \in [\![X_2]\!]\}$$

The set $\Omega(X_1, X_2)$ describes those propositional variables that are involved in a minimal difference between a model of X_1 and a model of X_2.

$$\Omega(X_1, X_2) = \bigcup \Delta^{min}(X_1, X_2)$$

In the symmetric difference operator Δ, information regarding the origins of the elements is lost. This loss is mitigated within the following non-commutative definition of symmetric difference: $\Delta_{\pm}(A, B) = (A \setminus B, B \setminus A)$. For any set of sentences X_1, X_2 define

$$\Delta_{\pm}(X_1, X_2) = \{(\Im \setminus \Im, \Im \setminus \Im) \mid \Im \in [\![X_1]\!], \Im \in [\![X_2]\!]\}$$

Now consider a subset of these sets which are minimal w.r.t. the cartesian product order $<_{\subseteq \times \subseteq}$ defined as usual by $(A, B) <_{\subseteq \times \subseteq} (C, D)$ iff $A \subseteq C$ and $B \subseteq D$.

$$\Delta^{min}_{\pm}(X_1, X_2) = min_{<_{\subseteq \times \subseteq}}(\Delta_{\pm}(X_1, X_2))$$

The adapted operator for $\Omega(\cdot, \cdot)$ is denoted $\Omega_{\pm}(X_1, X_2)$ which is defined as the pair of two sets: the first (second) argument consists of all propositional variables contained in the first (second) argument of some pair in $\Delta^{min}_{\pm}(X_1, X_2)$.

$$\Omega_{\pm}(X_1, X_2) = (\bigcup\{Y_1 \mid \exists Y_2 : (Y_1, Y_2) \in \Delta^{min}_{\pm}(X_1, X_2)\}, \bigcup\{Y_2 \mid \exists Y_1 : (Y_1, Y_2) \in \Delta^{min}_{\pm}(X_1, X_2)\})$$

The Satoh revision operator $*_S$ [34] defines the models of the revision result as those models of the trigger for which there is a model of the knowledge base with minimal symmetric difference.

$$[\![B *_S \alpha]\!] = \{\mathfrak{I} \in [\![\alpha]\!] \mid \text{There is } \mathfrak{J} \in [\![B]\!] \text{ with } \mathfrak{I}\Delta\mathfrak{J} \in \Delta^{min}(B, \alpha)\}$$

A natural weakening of this operator is the following one, which is coined *weak Satoh revision* here.

$$[\![B *_{wkS} \alpha]\!] = \{\mathfrak{I} \in [\![\alpha]\!] \mid \text{There is } \mathfrak{J} \in [\![B]\!] \text{ with } \mathfrak{I}\Delta_{\pm}\mathfrak{J} \in \Delta^{min}_{\pm}(B, \alpha)\}$$

Note that in both Satoh revision operators, minimality concerns the whole set of models of the knowledge base. In contrast to this, the operator of Borgida [5] considers for each model of the knowledge base the models of the trigger that are minimally distant. Borgida revision is defined as follows: If $B \cup \{\alpha\}$ is consistent, then $[\![B *_B \alpha]\!] = [\![B \cup \{\alpha\}]\!]$. Otherwise,

$$[\![B *_B \alpha]\!] = \bigcup_{\mathfrak{J} \in [\![B]\!]} \{\mathfrak{I} \in [\![\alpha]\!] \mid \mathfrak{I}\Delta\mathfrak{J} \in \Delta^{min}(\mathfrak{J}, \{\alpha\})\}$$

The operator of Weber [35] puts all those models of the trigger into the revision result for which there is a model of the knowledge base that differs at most in the propositional variables involved in a minimal difference. In the trivial case where $B \cup \{\alpha\}$ is consistent the definition is $[\![B *_B \alpha]\!] = [\![B \cup \{\alpha\}]\!]$. Else:

$$[\![B *_W \alpha]\!] = \{\mathfrak{I} \in [\![\alpha]\!] \mid \text{There is } \mathfrak{J} \in [\![B]\!] \text{ s.t. } \mathfrak{I} \setminus \Omega(B, \alpha) = \mathfrak{J} \setminus \Omega(B, \alpha)\}$$

A natural variant of the Weber operator uses the non-commutative definition of symmetric difference. In lack of a better name this operator is coined *weak Weber* operator.

$$[\![B *_{wkW} \alpha]\!] = \{\mathfrak{I} \in [\![\alpha]\!] \mid \text{There is } \mathfrak{J} \in [\![B]\!] \text{ s.t. } \mathfrak{I} \setminus pr_1(\Omega_{\pm}(B, \alpha)) = \mathfrak{J} \setminus pr_2(\Omega_{\pm}(B, \alpha))\}$$

7.6 Equivalence Results

The first equivalence result states that Satoh revision can be represented by disjunctively closed bi-implication-based reinterpretation operators $\circ^{\leftrightarrow}_{\gamma}$. The proof uses the known fact (re-stated below) that Satoh revision is nothing else than skeptical DS revision.

Theorem 7.17 ([7], Corollary 4.8) *Skeptical DS revision is Satoh revision:*

$$[\![B *_{DS} \alpha]\!] = [\![B *_S \alpha]\!]$$

For skeptical DS revision, due to its construction with bridging axioms, it can be shown that it is represented by an operator $\circ_{\gamma}^{\overleftrightarrow{}}$ with a simple selection function γ.

Theorem 7.18 *Skeptical DS revision $*_{DS}$ can be represented by an operator $\circ_{\gamma}^{\overleftrightarrow{}}$ where γ is defined independently of B (and α). That is, there is a selection function γ such that for any KB B and formula α the following holds:*

$$[\![B *_{DS} \alpha]\!] = [\![Cn^{\mathcal{P}}(B \circ_{\gamma}^{\overleftrightarrow{}} \alpha)]\!]$$

Here the function $\gamma = \gamma_1$ is defined as

$$\gamma_1(H) = \{\ X \in H \mid X \cap Bimpl \text{ is maximal in } \{X' \cap Bimpl \mid X' \in H\}\ \}$$

Proof See p. 167. □

As a corollary one gets the following theorem.

Theorem 7.19 *Satoh revision can be simulated by disjunctively closed bi-implication-based reinterpretation operators $\circ_{\gamma}^{\overleftrightarrow{}}$. That is, there is a selection function γ s.t. for any KB B and formula α such that:*

$$[\![B *_S \alpha]\!] = [\![Cn^{\mathcal{P}}(B \circ_{\gamma}^{\overleftrightarrow{}} \alpha)]\!]$$

Here the function $\gamma = \gamma_1$ is defined as

$$\gamma_1(H) = \{\ X \in H \mid X \cap Bimpl \text{ is maximal in } \{X' \cap Bimpl \mid X' \in H\}\ \}$$

In case of DS choice revision $*^c_{\mathrm{DS}}$ the outcome depends on a selection function c. Hence, one can mimic such an operator only by a reinterpretation operator that uses a maxi-choice AGM selection function (depending on c).

Proposition 7.20 *Let B a KB, α a trigger and c be a selection function for the bc extensions over B and α. There is a maxichoice selection function γ_c for Bimpl such that.*

$$[\![B *^c_{DS} \alpha]\!] = [\![Cn^{\mathcal{P}}(B \circ_{\gamma}^{\leftrightarrow} \alpha)]\!]$$

Proof See p. 179. □

Similar observations as for the theorems above also lead to the representation of implication-based skeptical DS revision and weak Satoh revision by disjunctively closed implication-based reinterpretation.

Theorem 7.21 *Implication-based skeptical DS revision and weak Satoh revision can be simulated by disjunctively closed implication-based reinterpretation operators*

$\circ_{\gamma}^{\rightarrow}$: *There is a selection function γ such that for any KB B and formula α it holds that*

$$[\![B *_{DS}^{\rightarrow} \alpha]\!] = [\![B *_{wkS} \alpha]\!] = [\![Cn^{\mathcal{P}}(B \circ_{\gamma}^{\rightarrow} \alpha)]\!]$$

Here the function $\gamma = \gamma_2$ is defined as

$$\gamma_2(H) = \{\ X \in H \mid X \cap \mathit{Impl} \text{ is maximal in } \{X' \cap \mathit{Impl} \mid X' \in H\}\ \}$$

Proof See p. 180 □

Also for the choice variant a representation by the corresponding reinterpretation operator can be proved.

Proposition 7.22 *Let B a KB, α a trigger and c be a selection function for the implication-based bc extensions over B and α. There is a maxichoice selection function γ_c for Impl such that:*

$$[\![B *_{DS}^{c,\rightarrow} \alpha]\!] = [\![Cn^{\mathcal{P}}(\mathit{Impl} *_{\gamma} (B' \cup \{\alpha\}))]\!]$$

Proof As above. □

For the proof of the theorem it is again sufficient to consider the proof for DS revision because it can be shown that skeptical implication-based revision $*_{\mathrm{DS}}^{\rightarrow}$ actually is the same as weak Satoh revision.

Theorem 7.23 $[\![B *_{DS}^{\rightarrow} \alpha]\!] = [\![B *_{wkS} \alpha]\!]$

Proof See p. 180. □

Weber revision is quite similar to Satoh, but it is more tolerant w.r.t. the models to be taken into account in the revision result. Dually, this tolerance w.r.t. the models means more skepticism regarding the sentences to keep in the revision result. Actually, this is reflected in the following theorem which says that Weber revision can be simulated by bi-implication-based reinterpretation: that is, in contrast to Satoh revision, the set of bi-implications is not exploited further w.r.t. logical consequences within the disjunctive closure. Moreover, as there is no additional closure of the bridging axioms, even full meet revision can be used, i.e., γ can be chosen as the identity function.

Theorem 7.24 *Weber revision can be represented by skeptical bi-implication-based reinterpretation $\circ^{Bimpl}$, i.e., for any KB B and formula α:*

$$[\![B *_{W} \alpha]\!] = [\![Cn^{\mathcal{P}}(B \circ^{\leftrightarrow} \alpha)]\!]$$

Proof See p. 180. □

With a similar argument one can show:

Theorem 7.25 *The weak Weber operator can be represented by skeptical implication-based reinterpretation.*

Proof See p. 181. □

Borgida revision is special in the sense that it considers the minimal symmetrical difference of models not globally, but locally for each model of the knowledge base. This model dependency can be simulated by literal-supported reinterpretation which allows the use of arbitrary primed literals as bridging axioms and thus allows the construction of arbitrary models of the knowledge base.

Theorem 7.26 *Borgida revision can be represented by literal-supported reinterpretation: there is γ for $Bimpl^+$ such that for any KB B and formula α:*

$$[\![B *_B \alpha]\!] = [\![Cn^{\mathcal{P}}(B \circ_\gamma^{lit} \alpha)]\!]$$

Proof See p. 181 □

7.7 Related Work

The reinterpretation framework uses bridging axioms to guide the resolution of conflicts. The most related work is that of Delgrande and Schaub [7]. But, additionally, there is a great deal of work in the general area of ontology change with which the reinterpretation framework presented in this chapter shares the main motivations. A classification of different forms of ontology-change operators with pointers to the literature is given by Flouris and colleagues [10]. Contributions to ontology mapping, ontology alignment, and mapping revision [22, 32] are related to my approach in so far as they investigate adequate constructions of semantic mappings, which are generalisations of bridging axioms. A more recent approach for mapping management in the paradigm of OBDA is given in [20].

Approaches for revising ontologies represented in description logics cannot rely on the general AGM strategy, as the consequence operator over DLs does not have the required properties (such as the deduction property) [11] so that different strategies are required. More recent approaches to ontology revision can be found in [33, 3].

The reinterpretation approach is symbol-oriented. A different symbol-oriented approach is described by Lang and Marquis [19]. Their revision operators are not based on bridging axioms but use the concept of forgetting [21]. As exemplified by the Weber operator [35] (see Sect. 7.5), there are strong connections between these approaches.

The notion of a prime implicate is used also in the approaches of [31, 36, 4]. In contrast to the approach of this chapter, all the approaches do not use prime implicates in the formulation of the postulates but (only) define new belief-revision operators based on prime implicates and show that they fulfil some classical postulates.

7.8 Summary

This chapter provided an investigation of reinterpretation operators and the induced knowledge-base operators from the representational perspective. Next to a classical representation theorem with postulates I considered also mutual simulations of operators, in particular showing, first, that operators with infinite revision results can be presented by operators with finite results and, second, that many classical revision operators can be simulated by reinterpretation operators. The latter is an argument in favour of the generality of reinterpretation: though all reinterpretation operators have been developed mainly for ontology-integration scenarios, they provide a sufficiently general framework for investigating classical belief-revision operators—at least, this has been shown for five classical belief-revision operators that are defined on purely semantical grounds.

Though the logic considered here was propositional logic, the symbol-oriented approach of reinterpretation lends itself for applications over knowledge bases represented in more expressive logics such as description logics or first-order logic (see, e.g., my contributions [24, 25, 9, 29, 28]). The general idea is to state bridging axioms about the relations of the predicate symbols and constants in the different name spaces by stating, e.g., the equivalence of the unary predicate symbol P' and P by $\forall x P'(x) \leftrightarrow P(x)$. Of course, the revision operators now become more complex. Moreover, it cannot be guaranteed that the conflicts can be solved by disambiguation—the knowledge bases of the sender and the receiver may be *reinterpretation incompatible* because they entail different cardinalities for their domains [25]. An open question is whether the representation theorems can be transferred to these logics. At least, the notion of uniform sets can also be defined for predicate logics and its fragments—though the prime implicate concept may not be purely semantical [26].

A further research line is a systematic study of reinterpretation operators w.r.t. the types of bridging axioms one is going to allow initially. As mentioned in the text above, one could consider bridging axioms such as $p' \leftrightarrow q$, which relate symbols hypothesised to be synonyms. Using a set H of such creative hypothesis may induce operators that are quite different from classical revision operators as the former may not be conservative: $B' \cup H$ may entail formulae $\beta \in sent(\mathcal{P})$ that do not already follow from B. Such creative behaviour does not occur for bridging axioms H chosen from *Impl* or *Bimpl*.

Also within this line of search one would set up criteria for genuine bridging axioms: For example, the literals used as bridging axioms for the representation of Borgida revision do not really bridge the meanings of symbols. In particular, the question arises whether for more restricted classes of bridging axioms a representation of Borgida revision is possible.

References

1. Alchourrón, C.E., Gärdenfors, P., Makinson, D.: On the logic of theory change: Partial meet contraction and revision functions. Journal of Symbolic Logic **50**, 510–530 (1985)
2. Armstrong, T., Marriott, K., Schachte, P., Søndergaard, H.: Two classes of boolean functions for dependency analysis. Science of Computer Programming **31**(1), 3–45 (1998)
3. Benferhat, S., Bouraoui, Z., Papini, O., Würbel, E.: A prioritized assertional-based revision for DL-Lite knowledge bases. In: E. Fermé, J. Leite (eds.) Proceedings of the 14th European Conference Logic in Artificial Intelligence (JELIA-14), *LNCS*, vol. 8761, pp. 442–456. Springer (2014). DOI 10.1007/978-3-319-11558-0_31
4. Bienvenu, M., Herzig, A., Qi, G.: Prime implicate-based belief revision operators. In: Proceedings of the 2008 Conference on ECAI 2008: 18th European Conference on Artificial Intelligence, pp. 741–742. IOS Press, Amsterdam, The Netherlands, The Netherlands (2008)
5. Borgida, A.: Language features for flexible handling of exceptions in information systems. ACM Transactions on Database Systems **10**(4), 565–603 (1985). DOI 10.1145/4879.4995
6. Dalal, M.: Investigations into a theory of knowledge base revision: Preliminary report. In: Proceedings of the 7th National Conference on Artificial Intelligence (AAAI-88), pp. 475–479. AAAI Press, St. Paul, Minnesota (1988)
7. Delgrande, J.P., Schaub, T.: A consistency-based approach for belief change. Artificial Intelligence **151**(1–2), 1–41 (2003)
8. Eiter, T., Gottlob, G.: On the complexity of propositional knowledge base revision, updates, and counterfactuals. Artificial Intelligence **57**, 227–270 (1992). DOI 10.1016/0004-3702(92)90018-S
9. Eschenbach, C., Özçep, Ö.L.: Ontology revision based on reinterpretation. Logic Journal of the IGPL **18**(4), 579–616 (2010). DOI doi:10.1093/jigpal/jzp039. First published online August 12, 2009
10. Flouris, G., Manakanatas, D., Kondylakis, H., Plexousakis, D., Antoniou, G.: Ontology change: Classification and survey. The Knowledge Engineering Review **23**(2), 117–152 (2008)
11. Flouris, G., Plexousakis, D., Antoniou, G.: On applying the AGM theory to dls and OWL. In: Y. Gil, E. Motta, V.R. Benjamins, M.A. Musen (eds.) Proceedings of the 4th International Semantic Web Conference (ISWC-05), *LNCS*, vol. 3729, pp. 216–231. Springer (2005). DOI 10.1007/11574620_18
12. Gärdenfors, P.: Rules for rational changes of belief. In: T. Pauli (ed.) Philosophical Essays Dedicated to Lennart Aquist on his Fiftieth Birthday, pp. 88–101. Philosophical Society and Department of Philosophy, Uppsala University (1982)
13. Gärdenfors, P.: Knowledge in Flux: Modeling the Dynamics of Epistemic States. The MIT Press, Bradford Books, Cambridge, MA (1988)
14. Hansson, S.O.: Belief base dynamics. Ph.D. thesis, Uppsala University, Uppsala (1991)
15. Hansson, S.O.: Reversing the Levi identity. Journal of Philosophical Logic **22**, 637–669 (1993)
16. Hansson, S.O.: A survey of non-prioritized belief revision. Erkenntnis **50**(2-3), 413–427 (1999)
17. Hansson, S.O.: A Textbook of Belief Dynamics. Kluwer Academic Publishers (1999)
18. Katsuno, H., Mendelzon, A.O.: Propositional knowledge base revision and minimal change. Artificial Intelligence **52**(3), 263–294 (1992)
19. Lang, J., Marquis, P.: Reasoning under inconsistency: A forgetting-based approach. Artificial Intelligence **174**(12-13), 799–823 (2010). DOI http://dx.doi.org/10.1016/j.artint.2010.04.023
20. Lembo, D., Mora, J., Rosati, R., Savo, D.F., Thorstensen, E.: Mapping analysis in ontology-based data access: Algorithms and complexity. In: M. Arenas, Ó. Corcho, E. Simperl, M. Strohmaier, M. d'Aquin, K. Srinivas, P.T. Groth, M. Dumontier, J. Heflin, K. Thirunarayan, S. Staab (eds.) Proceedings of the 14th International Semantic Web Conference (ISWC-15), Part I, *LNCS*, vol. 9366, pp. 217–234. Springer (2015). DOI 10.1007/978-3-319-25007-6_13
21. Lin, F., Reiter, R.: Forget it! In: Proceedings of the AAAI Fall Symposium on Relevance, pp. 154–159 (1994)
22. Meilicke, C., Stuckenschmidt, H.: Reasoning support for mapping revision. Journal of Logic and Computation (2009)

23. Noy, N.F.: Semantic integration: A survey of ontology-based approaches. SIGMOD Record **33**(4), 65–70 (2004)
24. Özçep, Ö.L.: Ontology revision through concept contraction. In: S. Artemov, R. Parikh (eds.) Proceedings of the Workshop on Rationality and Knowledge at the 18th European Summerschool in Logic, Language, and Information (ESSLLI-06), pp. 79–90 (2006)
25. Özçep, Ö.L.: Towards principles for ontology integration. In: C. Eschenbach, M. Grüninger (eds.) Proceedings of the 5th international conference on formal ontology in information systems (FOIS-08), vol. 183, pp. 137–150. IOS Press (2008)
26. Özçep, Ö.L.: Semantische integration durch reinterpretation - ein formales modell. Ph.D. thesis, Department for Informatics, MIN-Faculty, University of Hamburg (2009). URL `http://ediss.sub.uni-hamburg.de/volltexte/2010/4428/`. (In German)
27. Özçep, Ö.L.: Knowledge-base revision using implications as hypotheses. In: B. Glimm, A. Krüger (eds.) Proceedings of the 35th Annual German Conference on Artificial Intelligence (KI-12), LNCS, pp. 217–228. Springer Berlin Heidelberg (2012)
28. Özçep, Ö.L.: Iterated ontology revision by reinterpretation. In: G. Kern-Isberner, R. Wassermann (eds.) Proceedings of the 16th International Workshop on Non-Monotonic Reasoning (NMR-16), pp. 105–114 (2016)
29. Özçep, Ö.L.: Minimality postulates for ontology revision. In: C. Baral, J.P. Delgrande, F. Wolter (eds.) Proceedings of the 15th International Conference on Principles of Knowledge Representation and Reasoning (KR-16), pp. 589–592. AAAI Press (2016)
30. Özçep, Ö.L.: Belief revision with bridging axioms. In: V. Rus, Z. Markov (eds.) Proceedings of the 30th International Florida Artificial Intelligence Research Society Conference (FLAIRS-17), Marco Island, pp. 104–109. AAAI Press (2017)
31. Pagnucco, M.: Knowledge compilation for belief change. In: A. Sattar, B.h. Kang (eds.) Proceedings of the 19th Australian Joint Conference on Artificial Intelligence (AI-06), *LNCS*, vol. 4304, pp. 90–99. Springer Berlin / Heidelberg (2006)
32. Qi, G., Ji, Q., Haase, P.: A conflict-based operator for mapping revision. In: B.C. Grau, J. Horrocks, B. Motik, U. Sattler (eds.) Proceedings of the 22nd International Workshop on Description Logics (DL-09), *CEUR Workshop Proceedings*, vol. 477 (2009)
33. Ribeiro, M.M., Wassermann, R.: Minimal change in AGM revision for non-classical logics. In: C. Baral, G.D. Giacomo, T. Eiter (eds.) Proceedings of the 14th International Conference on Principles of Knowledge Representation and Reasoning (KR-14). AAAI Press (2014)
34. Satoh, K.: Nonmonotonic reasoning by minimal belief revision. In: Proceedings of the International Conference on Fifth Generation Computer Systems (FGCS-88), pp. 455–462. OHMSHA Ltd. Tokyo and Springer Verlag (1988)
35. Weber, A.: Updating propositional formulas. In: Proceedings of the 1st International Expert Database Systems Conference, pp. 487–500 (1986)
36. Zhuang, Z., Pagnucco, M., Meyer, T.: Implementing iterated belief change via prime implicates. In: M. Orgun, J. Thornton (eds.) Proceedings of the 20th Australian Joint Conference on Artificial Intelligence (AI-07), *LNCS*, vol. 4830, pp. 507–518. Springer Berlin / Heidelberg (2007)

Appendix

The following space saving abbreviations for $p \in \mathcal{P}$ are used throughout the proofs in this section: $\overleftrightarrow{p} = p \leftrightarrow p'$, $\overrightarrow{p} = p \rightarrow p'$ and $\overleftarrow{p} = p' \rightarrow p$. Remember also the abbreviations $Bimpl = \{\overleftrightarrow{p} \mid p \in \mathcal{P}\}$ and $Impl = \{\overrightarrow{p}, \overleftarrow{p} \mid p \in \mathcal{P}\}$.

Proof of Theorem 7.7

Let $(Im_i)_{i\in I}$ be the set of all implication-based consistent belief-change extensions for B and $\{\alpha\}$. First note that the maximality of the Im_i has the effect that for every $p \in \mathcal{P}$ at least one of $p \to p', p' \to p$ is contained in Im_i. Because, suppose that neither of $\overrightarrow{p}, \overleftarrow{p}$ is contained in Im_i. The maximality of Im_i entails $B' \cup Impl_i \cup \{\alpha\} \models \neg\overrightarrow{p} \wedge \neg\overleftarrow{p}$ and so $B' \cup Im_i \cup \{\alpha\} \models \bot$, which contradicts the fact that $B' \cup Im_i \cup \{\alpha\}$ is consistent.

Now I start the proof of the theorem by assuming that B is a formula in DNF. Let $c(I) = k$. I show that $B' \cup Im_k \cup \{\alpha\} \equiv_{\mathcal{P}} \lceil \mathcal{B} \rceil_k^{\to} \cup \{\alpha\}$ by proving the two implicit directions.

'Right to left': Let $B' \cup Im_k \cup \{\alpha\} \models \beta$ for $\beta \in sent(\mathcal{P})$. We have to show $\lceil \mathcal{B} \rceil_k^{\to} \cup \{\alpha\} \models \beta$. Let be given a model $\mathfrak{I} \models \lceil \mathcal{B} \rceil_k^{\to} \cup \{\alpha\}$. Then there is a dual clause cl in $\lceil \mathcal{B} \rceil_k^{\to}$ such that $\mathfrak{I} \models cl$. (Remember that a dual clause is just a conjunction of literals). For every literal li in cl one of the cases mentioned in Table 7.1 holds.

So there are 6 different types of literals in cl. This justifies the following representation of cl in $\lceil \mathcal{B} \rceil_k^{\to}$.

$$\begin{aligned} kl = {} & p_1^1 \wedge \cdots \wedge p_{n_1}^1 \wedge p_1^2 \wedge \cdots \wedge p_{n_2}^2 \wedge p_1^3 \wedge \cdots \wedge p_{n_3}^3 \\ & \wedge \neg p_1^4 \wedge \cdots \wedge \neg p_{n_4}^4 \wedge \neg p_1^5 \wedge \cdots \wedge \neg p_{n_5}^5 \\ & \wedge \neg p_1^6 \wedge \cdots \wedge \neg p_{n_6}^6 \end{aligned}$$

Define a new interpretation $\mathfrak{I}'$ in the following way:

- $\mathfrak{I}'(p_i'^1) = \mathfrak{I}'(p_i^1) = 1 = \mathfrak{I}(p_i^1)$;
- $\mathfrak{I}'(p_i'^2) = \mathfrak{I}'(p_i^2) = 1 = \mathfrak{I}(p_i^2)$;
- $\mathfrak{I}'(p_i'^3) = 0 \neq \mathfrak{I}(p_i^3) = 1$; $\mathfrak{I}'(p_i^3) = \mathfrak{I}(p_i^3) = 1$;
- $\mathfrak{I}'(p_i'^4) = \mathfrak{I}'(p_i^4) = 0 = \mathfrak{I}(p_i^4)$;
- $\mathfrak{I}'(p_i'^5) = \mathfrak{I}'(p_i^5) = 0 = \mathfrak{I}(p_i^5)$;
- $\mathfrak{I}'(p_i'^6) = 1 \neq \mathfrak{I}(p_i^6) = 0$; $\mathfrak{I}'(p_i^6) = \mathfrak{I}(p_i^6) = 0$;
- if r is a propositional symbol in $\mathcal{P}$ with $r \neq p_i^j$ and $r' \neq p_i'^j$, let $\mathfrak{I}'(r') = \mathfrak{I}(r)$;

From the construction of $\mathfrak{I}$ it follows that $\mathfrak{I}'_{\restriction\mathcal{P}} = \mathfrak{I}_{\restriction\mathcal{P}}$ and $\mathfrak{I}' \models B' \cup Im_k \cup \{\alpha\}$. So $\mathfrak{I}' \models \beta$ and hence $\mathfrak{I} \models \beta$.

'Left to right': Now suppose that $\lceil \mathcal{B} \rceil_k^{\to} \models \beta$ and let $\mathfrak{I} \models B' \cup Im_k \cup \{\alpha\}$. That is, there is a dual clause cl' in B' of the form

case	form in $\lceil \mathcal{B} \rceil_k^{\to}$	form in B	implications in $Impl_k$
I	p	p	$p' \to p, p \to p'$
II	p	p	$p' \to p$
III	p	$\neg p$	$p' \to p$
IV	$\neg p$	$\neg p$	$p' \to p, p \to p'$
V	$\neg p$	$\neg p$	$p \to p'$
VI	$\neg p$	p	$p \to p'$

Table 7.1 Cases for literals

$$p_1'^1 \wedge \dots \wedge p_{n_1}'^1 \wedge p_1'^2 \wedge \dots \wedge p_{n_2}'^2 \wedge \neg p_1'^3 \wedge \dots \wedge \neg p_{n_3}'^3$$
$$\wedge \neg p_1'^4 \wedge \dots \wedge \neg p_{n_4}'^4 \wedge \neg p_1'^5 \wedge \dots \wedge \neg p_{n_5}'^5$$
$$\wedge p_1'^6 \wedge \dots \wedge p_{n_6}'^6$$

It follows that $\Im(p_i^1) = \Im(p_i^2) = 1$ and $\Im(p_i^2) = \Im(p^5) = 0$. (Because of the types of the literals and the fact that the bridging axioms are made true.) Moreover, as $p_i^3 \rightarrow p_i'^3$ and $p_i'^6 \rightarrow p_i^6$ are not in Im_k, the maximality of Im_k entails $B' \cup Im_k \cup \{\alpha\} \models p_i^3 \wedge \neg p_i'^3$ and $B' \cup Im_k \cup \{\alpha\} \models \neg p_i^6 \wedge \neg p_i'^6$. Therefore we also have $\Im(p_i^3) = 1$ and $\Im(p_i^6) = 0$. Finally, this entails $\Im \models \lceil \mathcal{B} \rceil_k^{\rightarrow} \wedge \alpha$, hence $\Im \models \beta$.

Proof of Proposition 7.8

Every formula can be transformed into CNF. Therefore $clause(B) \equiv B$ and so it is sufficient to show $prime(B) \equiv clause(B)$. Clearly, $prime(B) \subseteq clause(B)$ and so trivially $clause(B) \models prime(B)$. In order to show $prime(B) \models clause(B)$ we show that for every clause $\beta \in clause(B)$ there is a $pr \in prime(B)$ fulfilling $pr \models \beta$. If β is a tautology, then $\emptyset \models \beta$. Therefore suppose that β is not tautological. If $\beta \in prime(B)$, then set $pr = \beta$. If $\beta \notin prime(B)$, there is a $\beta' \in clause(B)$, s.t. β' is a proper subclause of β. Because β is finite, it has only finitely many proper subclauses s.t. a minimal subclause β' can be chosen. There is a $\beta' \in clause(B)$ with: β' is a proper subclause of β and there is no proper subclause $\beta'' \in clause(B)$ of β'. In the end, $\beta' \in prime(B)$.

Proof of Proposition 7.9

Assume for contradiction, e.g., $prime(B_1) \not\subseteq prime(B_2)$. (The other case is proved similarly.) Then there is a prime implicate pl of B_1 that is not a prime implicate with respect to B_2. But $B_2 \models pl$, so there must be a prime implicate $pl' \subsetneq pl$ of B_2. In particular $B_2 \models pl'$, but then also $B_1 \models pl'$, which results in the contradicting assertion that pl cannot be a prime implicate of B_1.

Proof of Proposition 7.13

Because $prime^{\mathcal{P}}(B\rho) \subseteq prime^{\mathcal{P} \cup \mathcal{P}'}(B\rho)$ it follows that for every $\beta \in sent(\mathcal{P})$: if $prime^{\mathcal{P}}(B\rho) \models \beta$, then $prime^{\mathcal{P} \cup \mathcal{P}'}(B\rho) \models \beta$. In order to show the other direction assume that $\beta \in sent(\mathcal{P})$ and $prime^{\mathcal{P}}(B\rho) \not\models \beta$. I introduce the following abbreviations: $\Gamma_A = prime^{\mathcal{P} \cup \mathcal{P}'}(B\rho)$ and $\Gamma_B = prime^{\mathcal{P}}(B\rho)$. It needs to be shown that $prime^{\mathcal{P} \cup \mathcal{P}'}(B\rho) \not\models \beta$. There is a model $\Im \models \Gamma_B \cup \{\neg\beta\}$. So one has to show that there is a model $\Im'$ of $\Gamma_A \cup \{\neg\beta\}$, too. The intended model can be constructed inductively by constructing interpretations $\Im_i$ such that:

$$\begin{aligned}
\Im = \Im_0 &\models \Gamma_B \cup \{\neg\beta\} \\
\Im_1 &\models \mathit{prime}^{\mathcal{P}\cup\{p'_1\}}(B\rho) \cup \{\neg\beta\} \\
&\ldots \\
\Im' = \Im_n &\models \mathit{prime}^{\mathcal{P}\cup\{p'_1,\ldots,p'_n\}}(B\rho) \cup \{\neg\beta\} \\
&= \Gamma_A \cup \{\neg\beta\}
\end{aligned}$$

The interpretation $\Im_i$ is constructed from $\Im_{i-1}$ just by modifying only the interpretation of p'_i in a minimal way. Let X denote all prime consequences in the set $\mathit{prime}^{\mathcal{P}\cup\{p'_1,\ldots,p'_i\}}(B\rho)$ that do contain p'_i at most positively. If $\Im_{i-1}(p'_i) = 1$, then $\Im_i(p'_i) = 1$. Otherwise $\Im_{i-1}(p'_i) = 0$. If there is an $\alpha \in X$ such that $\Im_{i-1} \models \neg\alpha$, then define $\Im_i(p'_i) = 1$. Else let $\Im_i(p'_i) = 0$. Clearly $\Im_{i-1} \models \neg\beta$ (as only the interpretation of p'_i may have changed). By definition $\Im_i \models X$. So the only thing to show is that $\Im_i \models pr$ for all prime implicates in $\mathit{prime}^{\mathcal{P}\cup\{p'_1,\ldots,p'_i\}}(B\rho)$ with a negative occurrence of p'_i. Let $pr = \neg p'_i \vee M$ where M is a disjunction of literals not containing p'_i. Assume that $\Im_i \models p'_i$, i.e., $\Im_i(p'_i) = 1$. I have to show $\Im_i \models M$. There are two cases: $\Im_{i-1}(p'_i) = 1$, then $\Im_{i-1} \models M$ and hence $\Im_i \models M$. Otherwise $\Im_{i-1}(p'_i) = 0$ and there is a $\alpha = p'_i \vee N \in X$ (where N denotes a disjunction of literals) such that $\Im_{i-1} \models \neg\alpha$, i.e., $\Im_{i-1} \models \neg p'_i \wedge \neg \bigwedge N$ and so $\Im_i \models \neg N$. Resolving α with pr gives the clause $cl = N \vee M$ that does not contain p'_i. So there is a prime clause pr'' in $\mathit{prime}^{\mathcal{P}\cup\{p'_1,\ldots,p'_i\}}(B\rho)$ that is a subclause of cl. But $\Im_{i-1} \models pr''$ and so $\Im_i \models pr''$. As $\Im_i \models \neg N$, one concludes $\Im_i \models M$.

Proof of Proposition 7.14

The proof rests on a lemma (see below), which I mention only in the context of this proof due to its technicality. The lemma refers to the function $g(\cdot,\cdot)$ which is defined in the following way: Let *im* be an implication of the form $\overrightarrow{p}$ or $\overleftarrow{p}$. Let α be a formula. If $im = \overrightarrow{p}$, then $g(im, \alpha)$ stands for the assertion "p occurs semantically negative or not at all in α". If $im = \overleftarrow{p}$, then $g(im, \alpha)$ stands for the assertion "p occurs semantically positive in α or not at all". An occurrence is semantically positive (negative, resp.) in α iff for all interpretations $\Im$: if $\Im_{[p/0]} \models \alpha$ ($\Im_{[p/1]} \models \alpha$, resp.), then $\Im_{[p/1]} \models \alpha$ ($\Im_{[p/0]} \models \alpha$, resp.).

Lemma. *Let $S = \{p_1, \ldots, p_n\}$ and let $S_n = \mathcal{P} \cup \mathcal{P}' \setminus \{p'_1, \ldots, p'_n\}$. Let $U \subseteq S$ be the symbols $p_i \in S$, such that $\{\overrightarrow{p_i}, \overleftarrow{p_i}\} \subseteq Im$. For all $p_i \in (S \cap symb(Im)) \setminus U$ let $im(p_i)$ denote the implication (either $\overrightarrow{p_i}$ or $\overleftarrow{p_i}$) contained in Im. Let $Z = clause^{S_n}(B\rho \cup Im)$ for short. Then:*

$$Z = \{\beta \in clause^{S_n}(B) \mid \textit{There is a clause } \epsilon \textit{ with: } \epsilon \in clause^{S_n}(B\rho \cup Im) \\ \textit{and } \epsilon \models \beta \textit{ and } \epsilon \textit{ does not contain any symbol of } S \setminus symb(Im) \\ \textit{and for all } p_i \in (S \cap symb(Im)) \setminus U\textit{: } g(im(p_i), \epsilon)\}$$

Proof It may be assumed that for all implications in *Im* there is no implication of the other direction, so $U = \emptyset$. Let $Im = \{im_1, \ldots, im_k\}$. Proof of $\supseteq$: Let $\beta \in clause^{S_n}(B)$ and let ϵ be a clause, s.t.: $\epsilon \in clause^{S_n}(B\rho \cup Im)$, $\epsilon \models \beta$, ϵ has no symbol in $\{p_{k+1}, \ldots, p_n\}$ and for $1 \leq i \leq k$ it holds that $g(ba(p_i), \epsilon)$. Hence $\beta \in clause^{S_n}(B\rho \cup Im)$.

Proof of $\subseteq$: Let $\beta \in clause^{S_n}(B\rho \cup Im)$. Because $\beta \in sent(S_n)$, $(B\rho\rho^{-1} \cup (Im)\rho^{-1} \models \beta$ follows, so $B \models \beta$; hence $\beta \in clause^{S_n}(B)$. It has to be shown that an $\epsilon \in clause^{S_n}(B\rho \cup Im)$ exists that fulfils the mentioned conditions. Let $\widetilde{Im}$ be an equivalent CNF of *Im* and let $\tilde{B}$ be an equivalent CNF of *B* and let $(\widetilde{B\rho \cup Im})$ be the formula $\tilde{B}\rho \wedge \widetilde{Im}$. Assume β has the form $\beta = (li_1 \vee \cdots \vee li_q)$. Because $(B\rho \tilde{\cup} Im) \models \beta$, $(B\rho \tilde{\cup} Im) \cup \{\neg\beta\}$ is inconsistent. So $\tilde{B} \wedge \widetilde{Im} \wedge \neg li_1 \wedge \cdots \wedge \neg li_q$ can be resolved to the empty clause.

If β is already the clause ϵ which fulfils the desired conditions, then set $\epsilon = \beta$. Else β contains a symbol p for which (i) $p \in \{p_{k+1}, \ldots, p_n\}$ or there is i, $1 \leq i \leq k$, s.t. $p = p_i$ and not $g(im_i, \beta)$. I call such a symbol p a bad symbol. Let r denote the number of bad symbols in β. By induction on the number j of bad symbols one can construct a sequence $(\beta_j)_{0 \leq j \leq r}$ of clauses $\beta_j \in clause(B\rho \cup Im)$ such that:

$$B\rho \cup Im \models \beta_r \models \ldots \models \beta_1 \models \beta_0 = \beta$$

and every β_j has exactly $r - j$ bad symbols; in particular, β_r has no bad symbols so that β_r is the desired ϵ. Assume that β_j are already constructed and in particular assume $B\rho \cup Im \models \beta_j$. Let p be a bad symbol of β_j. W.l.o.g it may be assumed that β_j is not a tautology. I first consider the case that $p \in \{p_{k+1}, \ldots, p_n\}$. No literal $\neg li_j$ containing p, can be resolved with $\tilde{B}\rho \wedge \widetilde{Im}$; resolving $\neg li_j$ with a complementary clause in $(\neg li_1 \wedge \cdots \wedge \neg li_q)$ would be possible only if β_j were a tautology. Similarly clauses with p are not used for the derivation of the empty clause. So there is a clause β_{j+1}, which is obtained from β_j by eliminating literals containing p and for which $B\rho \cup Im \models \beta_{j+1}$ and $\beta_{j+1} \models \beta_j$. Moreover β_{j+1} has exactly $r - j - 1$ bad symbols.

In the second case β_j contains a symbol p_i, $1 \leq i \leq k$ for which $g(im_i, \beta_j)$ does not hold. W.l.o.g. assume $im_i = p_i \rightarrow p_i'$. So β_j does not contain p semantically negative. In particular β_j contains a literal li_j that contains p_i syntactically positively. Again $B\rho \cup Im \cup \{\neg\beta_j\}$ is inconsistent and so a derivation of the empty clause exists. The clause $\neg li_j$ contains p_i negatively. It cannot resolve with a clause in $(\tilde{B}\rho \wedge \widetilde{Im})$. A resolution with a clause in $\neg li_1 \wedge \cdots \wedge \neg li_q$ is not possible either—otherwise β_j would be a tautology. The clause β_{j+1} is obtained from β_j by removing the literal p_i. Again $\beta_{j+1} \models \beta$ and $B\rho \cup Im \models \beta_{j+1}$ and β_{j+1} $r - j - 1$ bad symbols. □

Now to the proof of the proposition. Let $pr \in prime^{\mathcal{P}}(B\rho \cup Im)$. Then $pr \in clause^{\mathcal{P}}(B)$. It has to be shown that $pr \in prime^{\mathcal{P}}(B)$. Assume that not

$pr \in prime^{\mathcal{P}}(B)$. That would mean that there is a clause $cl \in clause^{\mathcal{P}}(B)$ that is a proper subclause of pr. There are two cases: (a) $cl \in clause^{\mathcal{P}\cup\mathcal{P}'}(B\rho \cup Im)$. (b) $cl \notin clause^{\mathcal{P}\cup\mathcal{P}'}(B\rho \cup Im)$. Both cases result in a contradiction. Case (a) contradicts the fact that pr is prime with respect to $(B\rho \cup Im)$. In case (b) it holds that $B\rho \cup Im \not\models cl$ and $cl \models pr$. The first assertion and the lemma entail the fact that cl contains a symbol p (i) for which no bridging axioms is contained in Im or (ii) for which a bridging axiom is contained in the false direction.

Case (i): The lemma entails that there is a clause cl' such that $cl' \in clause^{\mathcal{P}}(B\rho \cup Im)$; p does not occur in cl' and $cl' \models pr$. Let pr' be a clause resulting from pr by removing all literals containing p. Then $B\rho \cup Im \models cl' \models pr'$. But this contradicts the primeness of pr w.r.t. $B\rho \cup Im$.

Case (ii): W.l.o.g. assume that $p' \rightarrow p \in Im$. Then cl contains a syntactically negative occurrence of p. Because of the lemma there is a clause $cl' \in clause^{\mathcal{P}}(B\rho \cup Im)$ such that cl' contains p only positively and $B\rho \cup Im \models cl' \models pr$. The symbol p can occur in pr at most positively. Otherwise, it would be the case that the clause pr', which results from pr by eliminating all literals $\neg p$, is entailed by $B\rho \cup Im$—contradicting the primeness of pr w.r.t. $B\rho \cup Im$. But as $cl \models pr$, also $cl[p/\bot] \models pr[p/\bot]$. As p occurs syntactically negative in cl, $cl[s/\bot]$ is a tautology; but then $pr[s/\bot]$ is a tautology, too—contradicting the primeness of pr w.r.t. $B\rho \cup Im$.

Proof of Theorem 7.15

Due to Proposition 7.9 it holds that $B' \cup Im \equiv_{\mathcal{P}\cup\mathcal{P}'} prime^{\mathcal{P}\cup\mathcal{P}'}(B' \cup Im)$. Now, in order to use Proposition 7.13 one has to represent $(B' \cup Im)$ as a set $B_1\rho$. The problem is that ρ will substitute all occurrences of the same symbol in B_1, so one cannot set $B_1 = B \cup Im$, as then also the non-primed symbols of Im would be substituted. So I proceed in the following way: For all symbols s in B I take a completely new symbol s''. Consider substitutions $\tau_1(s) = s''$, $\tau_2(s') = s''$ for symbols s in B. Now define the set $B_1 = B\tau_1 \cup Im\tau_2$. Let ρ be the substitution such that any s'' is substituted by s'. Then $B_1\rho = B' \cup Im$. Now because of Proposition 7.13 one gets $prime^{\mathcal{P}\cup\mathcal{P}'}(B_1\rho) \equiv_{\mathcal{P}} prime^{\mathcal{P}}(B_1\rho)$. But $prime^{\mathcal{P}}(B_1\rho)$ is $prime^{\mathcal{P}}(B' \cup Im)$ and according to Proposition 7.14 this is a subset of $prime^{\mathcal{P}}(B)$. Hence I set $X = prime^{\mathcal{P}}(B' \cup Im)$ which is easily seen to be a uniform set.

Proof of Theorem 7.16

'Left to right': Let B, α be given. Clearly $*^{c,\rightarrow}_{DS}$ fulfils (R1) and (R2). Let Im_k denote the set of implications underlying the bc extension chosen by c and let H_k be the set of prime implicates corresponding to Im_k according to Theorem 7.15. The fulfilment of (R3) follows by letting $H = \{H_k\}$. (R4) is fulfilled because $B \dotplus_k \alpha \models H_k$ and for all other uniform sets H the maximality of H_k implies $B \dotplus_k \alpha \models \neg \bigwedge H$. (R5) holds because if α and β are consistent with the same set of uniform sets, they are consistent with same set of implications. The definition of selection function guarantees that for bc-extensions w.r.t. α and w.r.t. β the same set of consistent implications and thus the same uniform set is entailed.

'Right to left': Let B, α be given. Let $(Impl_i)_{i \in I}$ be the set of bc extensions to the given bc scenario. I show, there is a selection function c s.t. $B * \alpha \equiv B *_{\text{DS}}^{c,\rightarrow} \alpha$. It can be assumed that B, α is consistent. According to (R3') there is $H \in U^{Impl}(B)$ such that $B * \alpha \equiv H \wedge \alpha$ or $B * \alpha \equiv H$. As (R2) is fulfilled, $B * \alpha \models \alpha$ and so $B * \alpha \equiv H \wedge \alpha$. The set of implications $(Impl_i)_{i \in I}$ induces a set $(H_i)_{i \in I}$ of uniform sets w.r.t. B. This follows from Theorem 7.15. Because $B * \alpha$ is consistent (according to (R1)) it follows that $\bigwedge H \wedge \alpha$ is consistent. Hence there is a H_k such that $H \subseteq H_k$, because all H_i are maximal uniform sets consistent with α. Because of tenacity $B * \alpha \models H_k$ or $B * \alpha \models \neg \bigwedge H_k$. But in the last case one would have $H \wedge \alpha \models \neg \wedge H_k$ or equivalently $\bigwedge H \wedge \bigwedge H_k \models \neg\alpha$ or equivalently $H_k \models \neg\alpha$, contradicting the consistency of H_k with α. Therefore $B * \alpha \models \bigwedge H_k \wedge \alpha$ and $\bigwedge H_k \wedge \alpha \models B * \alpha$. So one can set $c(I) = k$. Then $B * \alpha \equiv B *_{\text{DS}}^{c,\rightarrow} \alpha$. Now if β is such that it leads with B to the same set $(Impl_i)_{i \in I}$ of bc extensions, then one has to guarantee that one chooses again $Impl_k$. Here comes uniformity to the rescue: The set of uniform sets w.r.t. B that are consistent with $B * \beta$ and the set of uniform sets consistent with $B * \alpha$ are the same. Therefore the logical uniformity postulate (R5) entails that the same H_k is chosen.

Proof of Theorem 7.18

I define γ as follows:

$$\gamma(H) = \{\ X \in H \mid X \cap Bimpl \text{ is maximal in } \{X' \cap Bimpl \mid X' \in H\}\ \}$$

γ selects from H those sets for which the intersection with the set of bi-implications *Bimpl* is maximal. Note, that this definition is completely independent of B, and hence the content of the theorem is stronger than to say that for any B one may define a selection function γ such that the representation holds. Let $(Bim_j^{\vee})_{j \in J}$ be the family of sets in $\gamma(\overline{Bimpl} \top (B' \cup \{\alpha\}))$. Because of the definition of the disjunctive closure and of the remainders it holds that for all $i \in I$ there is a $j \in J$ s.t. $Bim_i \subseteq Bim_j^{\vee}$, $Bim_j^{\vee} \cap Bimpl = Bim_i$ and

$$Bim_j^{\vee} \subseteq Cn(Bim_i) \tag{7.1}$$

Conversely, because of the definition of γ one has for every $j \in J$ an $i \in I$ such that

$$Bim_j^{\vee} \supseteq Bim_i \tag{7.2}$$

Proof of $B *_{DS} \alpha \supseteq Cn^{\mathcal{P}}(\overline{Bimpl} *_\gamma (B' \cup \{\alpha\}))$: Let $\beta \in Cn(\mathcal{P})\overline{Bimpl} *_\gamma (B' \cup \{\alpha\})$, i.e., $\beta \in sent(\mathcal{P})$ and $(\bigcap_{j \in J} Bim_j^{\vee}) \cup B' \cup \{\alpha\} \models \beta$. So, for all $j \in J$ it holds that $Bim_i^{\vee} \cup B' \cup \{\alpha\} \models \beta$ and so $Bim_j^{\vee} \models (\bigwedge B' \wedge \alpha) \rightarrow \beta$. Together with (7.1) it follows that for all $i \in I$: $Bim_i \models (\bigwedge B' \wedge \alpha) \rightarrow \beta$, hence $(\bigwedge B' \wedge \alpha) \rightarrow \beta \in Cn(Bim_i) \subseteq Cn(Bim_i \cup B' \cup \{\alpha\}) = E_i$ for all $i \in I$. Consequently, $\beta \in E_i$ for all $i \in I$ and lastly $\beta \in \bigcap_{i \in I} E_i = B *_{DS} \alpha$.

Proof of $B *_{DS} \alpha \subseteq Cn^{\mathcal{P}}(\overline{Bimpl} *_\gamma (B' \cup \{\alpha\}))$: Let $\beta \in B *_{DS} \alpha = \bigcap_{i \in I} E_i$, i.e. $\beta \in sent(\mathcal{P})$, and for all $i \in I$: $Bim_i \cup B' \cup \{\alpha\} \models \beta$ and hence $Bim_i \models (\bigwedge B' \wedge \alpha) \rightarrow \beta$. Because of the compactness property of propositional logic one has for every $i \in I$ a finite subset $Bim_i^f \subseteq Bim_i$ such that $Bim_i^f \models (\bigwedge B' \wedge \alpha) \rightarrow \beta$. Because B is finite, so is the set I, which is the index set of all bc extensions E_i. Let $I = \{1, \ldots, k\}$. There are only finitely many maximal sets of bridging axioms Bim_i and finitely many extensions E_i. So the disjunction $\bigvee_{i \in I} Bim_i^f$ is defined and the following holds:

$$\bigvee_{i \in I} Bim_i^f \models (\bigwedge B' \wedge \alpha) \rightarrow \beta \tag{7.3}$$

For all $i \in I$ let $n_i = |Bim_i|$ be the number of elements in Bim_i^f and $N_i = \{1, \ldots, n_i\}$. Every set Bim_i, $i \in I$, is representable as $Bim_i \equiv \bigwedge_{j=1}^{n_i} (p_{ij} \leftrightarrow p'_{ij})$. Applying the distribution law $\bigvee_{i \in I} Bim_i^f$ is transformable in a conjunction of disjunctions of bi-implications:

$$\bigvee_{i \in I} Bim_i^f \equiv \bigwedge_{(j_1,\ldots,j_k) \in N_1 \times \cdots \times N_k} \bigvee_i^k (p_{j_i} \leftrightarrow p'_{j_i}) \tag{7.4}$$

Because of (7.2) for every $j \in J$ there is an $i \in I$ with $Bim_j^{\vee} \supseteq Bim_i$. Now for every $(j_1, \ldots, j_k) \in N_1 \times \cdots \times N_k$ it holds that $(p_{j_i} \leftrightarrow p'_{j_i}) \in Bim_i$ and hence for every $(j_1, \ldots, j_k) \in N_1 \times \cdots \times N_k$ also $\bigvee_i^k (p_{j_i} \leftrightarrow p'_{j_i}) \in Bim_j^{\vee}$ holds. Hence for every $j \in J$ it is true that $Bim_j^{\vee} \models \bigwedge_{(j_1,\ldots,j_k) \in N_1 \times \cdots \times N_k} \bigvee_i^k (p_{j_i} \leftrightarrow p'_{j_i})$. With (7.4) it follows that for every $j \in J$ that $Bim_j^{\vee} \models \bigvee_{i \in I} Bim_i^f$ and with the entailment in (7.3) it further follows that $Bim_j^{\vee} \models (\bigwedge B' \wedge \alpha) \rightarrow \beta$. In the end: $\beta \in Cn^{\mathcal{P}}(\overline{Bimpl} *_\gamma (B' \cup \{\alpha\}))$.

Proof of Proposition 7.20

Let $(E_i)_{i \in I}$ be the set of bc extensions for B and α and let $(EQ_i)_{i \in I}$ be the set of bi-implications on which the E_i are based. Let γ be defined by $\gamma(EQ \top (B' \cup \{\alpha\})) = \{EQ_i\}$ for $i = c(I)$. Then by definition: $EQ *_\gamma (B' \cup \{\alpha\}) = \bigcap (\gamma(EQ \top (B' \cup \{\alpha\}))) \cup B' \cup \{\alpha\} = EQ_i \cup B' \cup \{\alpha\}$. Hence $Cn^{\mathcal{P}}(EQ *_\gamma (B' \cup \{\alpha\})) = Cn(EQ *_\gamma (B' \cup \{\alpha\})) \cap sent(\mathcal{P}) = E_i = B \dotplus_c \alpha$.

Proof of Proposition 7.21

The proof for the equivalent representation of $*_{\mathrm{DS}}^{\rightarrow}$ by a disjunctively closed implication-based reinterpretation operator uses the same construction as in the proof of Theorem 7.18, as the construction does not use the special property of bi-implications. With this also the representation for weak Satoh follows due to Theorem 7.23.

Proof of Theorem 7.23

With Theorem 7.7 Theorem 7.23 is an immediate corollary: Assume that b is given in complete disjunctive normal form. $\Im \models \lceil B \rceil^{\rightarrow} \wedge \alpha$ iff $\Im \models \lceil B \rceil_i^{\rightarrow}$ for some i. Now, in $\lceil B \rceil_i^{\rightarrow}$ all propositional variables p for which either $\overrightarrow{p}$ or $\overleftarrow{p}$ does not occur in Im_i all associated occurrences in the dual clauses (which correspond actually to models) are flipped into the other polarity so that all dual clauses have for each occurrence of p the same polarity. But this means that there is a model $\Im \models B$ which differs from $\Im$ exactly in the corresponding polarities for the ps with missing bridging axioms in Im_i; this is the same as saying that $\Im \Delta_{\pm} \Im \in \Delta_{\pm}^{min}(B, \alpha)$.

Proof of Theorem 7.24

For the proof of this theorem I use an alternative characterization of Weber revision with the forgetting operator Θ_S (see Chap. 2).

$$[\![B *_W \alpha]\!] = [\![\Theta_{\Omega(B,\alpha)}(B) \wedge \alpha]\!]$$

According to definition $B \circ^{Bimpl} \alpha = \underbrace{\bigcap (Bimpl \top (B' \cup \alpha))}_{=:X} \cup B' \cup \{\alpha\}$. Due to interpolation it holds that $Cn^{\mathcal{P}}(X \cup B' \cup \alpha) = Cn^{\mathcal{P}}(Cn^{\mathcal{P}}(X \cup B') \cup \alpha)$. Now one can verify that

$$Cn^{\mathcal{P}}(X \cup B') = Cn^{\mathcal{P}}(\Theta_{\{p \in \mathcal{P} \mid \overleftrightarrow{p} \notin X\}}(B))$$

(Because, for example, $\Theta_{p',q'}(B' \cup p' \leftrightarrow p)) = (B'[p'/1, q'/0] \wedge p) \vee (B'[p'/1, q'/1] \wedge p) \vee (B'[p'/0, q'/0] \wedge \neg p) \vee (B'[p'/0, q'/1] \wedge \neg p) \equiv_{\mathcal{P}} (B'[p'/p, q'/0] \vee B'[p'/p, q'/1] = \Theta_{\{q\}}(B)$.) Now $\{p \in \mathcal{P} \mid \overleftrightarrow{p} \notin X\} = \Omega(B, \alpha)$. Hence due to $[\![B *_W \alpha]\!] = [\![\Theta_{\Omega(B,\alpha)}(B) \cup \{\alpha\}]\!]$, the assertion follows.

Proof of Theorem 7.25

The proof works in the same way as the proof of Theorem 7.24. For this I use the alternative characterisation of weak Weber as

$$[\![B *_W \alpha]\!] = [\![\Theta_{pr_1(\Omega_\pm(B,\alpha)) \cup pr_2(\Omega_\pm(B,\alpha))}(B) \wedge \alpha]\!]$$

Proof of Theorem 7.26

The core of the proof is to find the correct definition of the selection function γ. Let $\Psi := \{p', \neg p' \mid p \in \mathcal{P}\}$.

$$\begin{aligned}\gamma(Z) = \{X \in Z \mid\ & X \cap \Psi \text{ is inclusion maximal within all } X' \cap \Psi \in Z \text{ and} \\ & X \cap \mathit{Bimpl} \text{ is inclusion maximal within all } X'' \cap \mathit{Bimpl} \\ & \text{for which there is } X''' \subseteq \mathit{Bimpl} \text{ with } X'' = (X \cap \Psi) \cup X'''\}\end{aligned}$$

This selection function first selects maximal sets of primed literals from Ψ. These maximal sets correspond just to the models of B'. Then it chooses maximal sets of the disjunctive closure of the bi-implications. But as was shown for the representation of Satoh revision by disjunctively closed bi-implications, this corresponds to considering minimal symmetrical difference.

Chapter 8
Conclusion

Abstract This final chapter gives an overall conclusion regarding the role of logics and representation theorems for computer science, sketches possible future work on automatising the process of finding representation theorems, and makes some suggestions on possible curricula in computer science.

Many Worlds and Representation

"A funny thing happened in the last few years. We began to lose the Closed World Assumption."[1] These words from a blog on data-science related issues summarises the situation of CS that motivated the research reported in this monograph. In various applications of CS, the designer will have to accept that his formal specification may have many models—some of which are possibly unintended. This is even true for the field of database systems, which is usually assumed to adhere to the closed-world assumption, in contrast to the field of logic, which adheres to the open-world assumption. A clear indication is the use of NULL values to handle the incompleteness of data stored in DBs. Though the NULL value construct is part of the SQL query standard, it is the most criticised one, mainly due to its non-intuitive semantics [6] which may easily lead to unintended DB models and hence also to unintended query answers.

Of course, one might say that the NULL value construct is a sign of a glitch in the SQL standard and that, in general, standardisations w.r.t. formal specifications can be understood as well-meant efforts to tame the multiplicity of (unintended) models. But unfortunately, even a totally worked-out standard cannot prevent the undesirable development of diseased misuse of formal specifications adhering to this standard. One example of such an unfortunate development can be termed "Importitis": instead of designing a formal specification fitting the requirements, the application designer

[1] Taken from: Kurt Cagle: Data Science and the open world assumption, blog entry, `http://www.datasciencecentral.com/profiles/blogs/data-modeling-and-the-open-world-assumption` (last access 14 July 2017)

Ö. L. Özçep, *Representation Theorems in Computer Science*,
https://doi.org/10.1007/978-3-030-25785-9_8

imports freely available formal specifications—in many cases without thinking about the consequences of imported axioms in this context. This in turn may easily lead to formal specifications without any model or with many unintended models.

Another example of such a development could be called "Annotationitis". The application designer uses formal specifications to annotate the entities of the domain without really asking what to represent. Everything gets labelled, but it is not clear to what ends. In this monograph I tried to argue for a methodological shift in the sense that the application designer should ask himself "What do I really want to represent, and does the formal specification really represent it?".

Having identified representation theorems as appropriate means for taming the multiplicity of possible models of formal specifications, this monograph worked out a representation framework and gave representation results for various CS applications. The definition of representation that I used appears to be appropriate for the applications considered in this monograph. One alleged weakness of my definition of representation is the semi-formality of the representation notion, in particular w.r.t. the condition that states that the representing structures must be "easily constructible". Of course, it is possible to formalise this term using some complexity theoretic notion. For example, one might state that a set of representing objects is easily constructible iff there exists a Turing Machine that enumerates all representing objects in some encoding using feasible time and space resources. But, actually, the informality of my representation definition is necessary due to the overall aim of supporting application designers in building only "intended models": there is the human component of "intentions" that make it hard to formalise "simplicity" on purely complexity theoretical terms without incorporating cognitive aspects. By the way, an analogous informality is inherent to the Church-Turing thesis stating that all "intuitively" calculable problems can be calculated by Turing machines.

Finding Representation Theorems Automatically

As explicated in the introduction, the overall goal of the research reported in this monograph is to support application designers in building good formal specifications for information systems. Inspecting the proofs of the representation results of this monograph in detail, it becomes clear that there is no obvious way to mechanise the process of representation. But in the end, this is what a company expects from a full support tool for the application engineer: a system that (semi-)automatically builds, analyses, possibly completes and corrects formal specifications and "overviews" the models of the formal specification by describing a class of representing models.

Of course, building for each individual information system its own support tool is a possible approach, but actually one is interested in general principles underlying the individual support tools, or in other words: one is interested in (principles of) a general-purpose support tool. And here machine learning (ML) enters the stage. The following paragraph contains a speculative outlook on how the representation framework developed in this monograph can be enhanced by ML methodology in order to build such a support tool.

Machine learning investigates general algorithms for finding hypotheses (models) fitting a sequence of training data and generalising to all data as well as possible. In the subfield of ML called Inductive Logic Programming (ILP) [7], the hypotheses are logical theories, mostly given as intensional rules, and the data in the sequence are basic (extensional) facts. So, ILP seems to be a good starting point for the task of generating formal specifications for an information system. The application designer trains the ML algorithm with simple facts expressing the intended behaviour of the system by, say, the allowed input-output instances, and the ML algorithm comes up with possible theories capturing this behaviour. Of course the ambitions must go further, as anticipated by the theoretical ambition underlying, e.g., Church's synthesis problem[2] [3] or the practical ambition of building a compiler that, given a specification in a high-level programming language, produces highly efficient code fulfilling the specification (see, e.g., [9]). In ILP terms, the aim is not only learning intensional rules for specifying the intended behaviour of a system but also learning possible designs of information systems producing the behaviour.

Clearly, this more ambitious task is not trivial, as this presupposes that one has first ideas on possible parameters or—in machine learning terms—first ideas on the possible hidden variables. Generally one has to come up with ideas on the relevant features that have to be incorporated into the model. In logical terms the hidden variables correspond to concepts and relations that are introduced to describe the architecture of the information system—and not just the input-output behaviour. So the problem is to determine the feature space, i.e., to extract correct features to specify data and hypotheses.

But, there is good news: recent ML research comes up with solutions as developed, e.g., in deep learning [8] that enable automatic feature extraction. Clearly, the problem of automatic representation is not solved by a straightforward application of ML methods: First of all, the logical setting is more demanding due to the general relational structure of the models, which, prima facie is more complicated than the simple vector-structure with feature-attributes. Secondly, representation in the sense of this monograph also requires identifying representing models and finding mappings from other models to the representing models. Hence, one has to extend the idea also to learning the set of potential representing models.

Though speculative, above thoughts on using ML methods for automatic representation with logical structures are not utopian because of the current trends and developments in ML. First of all, it has been clear to the deep learning community that finding correct features means finding the right representation—and so it is not a pure coincidence that "feature learning" is sometimes also called "representation learning" as in the title of [1]. So, whereas earlier ML research was already satisfied by the promising outcomes of ML algorithms, current ML research recognises the necessity of understanding those outcomes (semantically). But even more, recent research in the intersection of logic and machine learning—sometimes mentioned

[2] Given a formal specification $\phi(In, Out)$ for the input-output behavior find a concrete function F such that $\phi(In, F(In))$

under the buzz word "Data Science"—can be considered as a twofold attempt: understanding and describing the results of machine learning algorithms in a logical setting as done, e.g., in [5], and, vice versa, using logical constraints to guide or optimise machine learning algorithms, as done, e.g., in [4].

Logics for Computer Science Curricula

If it was not already clear before, this monograph might have convinced the reader of the pre-eminent role that logic plays for CS. Acknowledging the importance of logic, it is surprising to see that many academic curricula in CS pay barely attention to logic—let alone to a smooth integration of logic with the subfields of CS. Hence I would like to use the last paragraphs of the conclusion to make a plea for a complete restructuring of CS curricula under the motto "Logic, Logic, and Logic"[3].

All students—not only master students but even more importantly: bachelor students—should be trained as early as possible with logic and formal methods, not only to get acquainted with the results produced in the discipline of logic but also to get trained in necessary skills for formalising own ideas, domains and information systems, for building models and using them, and in general for learning to speak, think and argue precisely.

Classical themes such as syntax, semantics, calculi for propositional and first-order logic should be the starting points laying the ground for advanced topics of logic and for the treatise of other logics. For example, students should become acquainted with the general ideas of finite model theory, descriptive complexity, proof theory and with "computational" logics such as (existential) second order modal logics, datalog, description logics, dynamic logic, temporal logic, probabilistic logic—ideas and themes that are usually not taught in mathematics courses.

In the other subfields of CS that are part of the curriculum the role of logics has to be clearly worked out and, even more, logic has to be the main thread. For example, in a course for database systems one should stress the role of first-order logic as query language rather than practicing SQL. On the other hand, the conception of the logic courses should refer to concrete applications from CS such as those mentioned in the introduction of this monograph. Moreover, in order to strengthen the trust in logic, students should be trained first in logical and functional programming paradigms (with Prolog, answer set programming, DLV, Haskell, Lisp) before becoming acquainted with the object-oriented paradigm.

Also for non-CS courses that are part of the curriculum a clear reference to CS is required. Students should be guided to ask also in mathematics courses those kinds of questions that a computer scientist is interested in. For this purpose early programming exercises in non-CS courses may prove useful.

---

[3] This is actually the title of a book [2] by the logician J. Boolos.

References

1. Bengio, Y., Courville, A., Vincent, P.: Representation learning: A review and new perspectives. IEEE Transactions on Pattern Analysis and Machine Intelligence **35**(8), 1798–1828 (2013). DOI 10.1109/TPAMI.2013.50
2. Boolos, G., Jeffrey, R.: Logic, Logic, and Logic. Harvard University Press, Cambridge, MA and London (1999)
3. Church, A.: Application of recursive arithmetic to the problem of circuit synthesis. Summaries of the Summer Institute of Symbolic Logic **1**, 3–50 (1957)
4. Deng, J., Ding, N., Jia, Y., Frome, A., Murphy, K., Bengio, S., Li, Y., Neven, H., Adam, H.: Large-scale object classification using label relation graphs. In: D. Fleet, T. Pajdla, B. Schiele, T. Tuytelaars (eds.) Proceedings of the 13th European Conference on Computer Vision (ECCV-14), Part I, *LNCS*, vol. 8689, pp. 48–64. Springer International Publishing (2014)
5. Grohe, M., Ritzert, M.: Learning first-order definable concepts over structures of small degree. ArXiv e-prints (2017)
6. Libkin, L.: SQL's three-valued logic and certain answers. ACM Transations on Database Systems **41**(1), 1:1–1:28 (2016). DOI 10.1145/2877206
7. Muggleton, S.: Inductive logic programming. New Generation Computing **8**(4), 295–318 (1991). DOI 10.1007/BF03037089
8. Schmidhuber, J.: Deep Learning in Neural Networks: An Overview. ArXiv e-prints (2014)
9. Schwartz, J.T., Dewar, R.B.K., Dubinsky, E., Schonberg, E.: Programming with Sets - An Introduction to SETL. Texts and Monographs in Computer Science. Springer (1986). DOI 10.1007/978-1-4613-9575-1

in the United States
masters